KT-437-421

159 962

Pesticides – Developments, Impacts, and Controls

Pesticides – Developments, Impacts, and Controls

Edited by

Gerry Best

Clyde River Purification Board, East Kilbride, Glasgow, UK

Douglas Ruthven

Scottish Agricultural Science Agency, Edinburgh, UK

THE ROYAL
SOCIETY OF
CHEMISTRY

The Proceedings of the Analytical Division's Symposium of The Royal Society of Chemistry Annual Chemical Congress, held on 10–13 April 1995, at Heriot-Watt University, Edinburgh, UK.

Special Publication No. 174

ISBN 0-85404-785-9

A catalogue record for this book is available from the British Library

Published by The Royal Society of Chemistry,
Thomas Graham House, Science Park, Milton Road,
Cambridge CB4 4WF, UK

Printed by Bookcraft (Bath) Ltd.

Preface

"Ministers to look into sheep dip dangers" (The Scotsman 5/5/95), "Pesticides blamed for eye defects" (The Herald 1/5/95), "Fruit and vegetables tainted with chemical" (The Times 3/5/95). Just three of the headlines that appeared in the UK press at the time that this book was being prepared for publication. The wording reflects the way that pesticides are viewed most often by the media and the general public. They are generally perceived as dangerous chemicals that can and do cause harm. It is often forgotten that pesticides have saved the lives of millions of people throughout the world. For example, in India during 1933-35, there were 100 million cases of malaria reported each year with 700,000 deaths from the disease. By 1966, after widespread use of DDT against mosquito larvae, there were only 150,000 cases with 1,500 deaths. Pesticides not only protect and enhance our lives but also our food from attack by fungi and competing weeds, and our homes and environment. Pre-harvest crop losses from pests are estimated at about 35% world wide with additional losses of 20% during transport and storage despite all our efforts to protect this food (*Chem. Br.*, 27,7,646). Without the use of pesticides as protectants, yields even now would decrease by about 40% (*Chem. Br.*,27,**11**,1010).

There have been great advances in the nature of pesticides since the first examples were synthesised over 60 years ago. Of particular significance in recent years has been the move to produce substances that attack the target pest but do not adversely affect the environment. The variety and the chemical complexity of the substances available world-wide is enormous and there are currently over 450 active ingredients registered for use in the UK. Despite the need for accurate measurement of their environmental impacts, there are reliable and sensitive analytical methods for detecting only about half of them at trace levels. New substances arrive, such as the new systemic fungicide reported in Chemistry in Britain in June 1995 (*Chem. Br.,* 31,**6**,466). Regulatory authorities and legislators must keep abreast of developments to ensure that the use of pesticides does not threaten the health of users, food consumers and the environment, and that adequate controls are in place.

These three themes, developments, impacts and the control of pesticides, were explored in the Analytical Division's symposium at the Annual Chemical Congress of the Royal Society of Chemistry held at Heriot-Watt University, Edinburgh, in April 1995.

This book contains the written versions of the papers that were presented at the symposium and brings the reader up to date with the latest information about many important issues surrounding the use of pesticidal substances. The editors are grateful to the contributors, not only for giving excellent lectures at the symposium, but also for presenting their papers in good time and requiring the minimum of alteration.

Gerry Best
(Clyde River Purification Board)

Douglas Ruthven
(Scottish Agricultural Science Agency)

Contents

Contributors

Dr. Michael F. Wilson
Pesticides Group, Central Science Laboratory, MAFF, Hatching Green, Harpenden, Herts., AL5 2BD

Dr. Richard P. Garnett
Monsanto plc., Thames Tower, Burleys Way, Leicester, LE1 3TP

Dr. Robert M. Perrin
Zeneca Agrochemicals, Fernhurst, Haslemere, Surrey, GU27 3JE

Dr. Andrew J. Gilbert
Pesticides Group, Central Science Laboratory, MAFF, Hatching Green, Harpenden, Herts, AL5 2BD

Dr. Bernard P. Nutley
Health and Safety Executive, HPD-B3, Rose Court, 2 Southwark Bridge, London, SE1 9HF

Professor Gerald A. Matthews
International Pesticide Application Research Centre, Dept. of Biology, Imperial College of Science, Technology and Medicine, Silwood Park, Ascot, SL5 7PY

Dr. Peter Smith
Strathclyde Regional Chemist's Department, 64 Everard Drive, Glasgow, G21 1XG

Dr. Kenneth Hunter
Scottish Agricultural Science Agency, East Craigs, Edinburgh, EH12 8NJ

Dr. Robert H. Foy
Agriculture and Environmental Science Division, Department of Agriculture for Northern Ireland, Newforge Lane, Belfast, BT9 5PX

Dr. Peter A. Chave
National Rivers Authority, Rivers House, Waterside Drive, Aztec West, Bristol, BS12 3UD

Dr. Andrée D. Carter
Soil Survey and Land Research Centre, Cranfield University, Shardlow Hall, Shardlow, Derby, DE72 2GN

Dr. Brian T. Croll
Anglian Water Services Ltd., Compass House, Chivers Way, Histon, Cambridge, CB4 4ZY

Dr. Ian M. Davies
Scottish Office Agriculture and Fisheries Department, Marine Laboratory, P.O.Box 101, Victoria Road, Aberdeen, AB9 8DB

Mr. Terry Tooby
Pesticides Safety Directorate, Mallard House, Kings Pool, Peasholme Green, York, YO1 2PX

Mrs Melanie G. C. Quinn
Water Research Centre plc., Henley Road, Medmenham, Marlow, Bucks., SL7 2HD

Mr. John Seddon
BASIS (Registration) Ltd, 2 St.John Street, Ashbourne, Derbyshire, DE6 1GH

Dr. Michael Pearson
National Rivers Authority, Toxic and Persistent Substances National Centre, Kingfisher House, Goldhay Way, Orton Goldhay, Peterborough, PE2 5ZR

Monitoring and Adapting to the Changes in Pesticide Use Profiles That Occur in Response to Modern Pest Control and Environmental Requirements

Michael F. Wilson

PESTICIDES GROUP, CENTRAL SCIENCE LABORATORY (CSL), MINISTRY OF AGRICULTURE, FISHERIES AND FOOD (MAFF), HATCHING GREEN, HARPENDEN, HERTFORDSHIRE AL5 2BD, UK

1. INTRODUCTION

Science and technology has given us the ability to characterise objects at great distance, objects which are tiny and to search with confidence for traces of chemicals which are present at minute levels in complex mixtures. However, this ability is accompanied by an increase in the costs of necessary equipment and resources, costs which must be balanced against the needs for which the information is being generated.

Terrestrial telescopes, and most recently the Hubble Space telescope in orbit around the Earth, have given astronomer the ability to observe stars, galaxies and other objects deep into the universe. Archaeology, always viewed as the art of searching for small clues about the past, has also benefited from technological advances. Using Spacebourne Imaging Radar-C (SIR-C/X-SAR) scientists at the Jet Propulsion Laboratory, California identified the site of the 'lost' city of Ubar in southern Oman. The city existed from 2800 BC but was abandoned in about 300 AD when caravan routes changed. In 1992 and with equipment mounted on the Space Shuttle Endeavour, the city was pinpointed on a radar scan of some 5000 km². The ancient tracks leading to the city fortress show as red lines on the radar image and guided archaeologists to the site.

Both astronomy and archaeology have been compared to looking for a needle in a haystack. This parallel can be equally applied to trace analysis including the analysis of pesticides, their metabolites and breakdown products. There are 725 pesticides listed in the current edition of the Pesticides Manual[1] and whilst only a fraction of these are approved for use in the United Kingdom, the international trade in foods and commodities means that the analyst needs to call on sophisticated technology and a high degree of skills and experience to provide the regulators with data on the occurrence or non-occurrence of residues in diet and in the environment. Refinements in extraction, clean-up, separation and confirmatory techniques have conferred upon the analyst the ability to find that needle in the chemical haystack which is often the matrix in which it occurs. However, the technical ability to carry out determinations must be harnessed to the needs of the policies and practices governing pesticide use.

2. THE DRIVING FORCES

It is one of the challenges facing the analyst not only to develop sensitive, unequivocal and robust methods for the determination of chemicals such as pesticide residues in a wide range of matrices but to hone those methods to the changing demands being placed upon them by (a) changes in usage practices, (b) legislation and (c) in efficiently offering the regulators data which will ultimately provide valued information to monitor agricultural practice and consumer exposure.

2.1 Agricultural Practices

Data collated by the MAFF Pesticide Usage Survey, part of the Central Science Laboratory, and collected from England, Wales and Scotland in 1993 demonstrate that most pesticides (86 %) are applied to arable crops. Of the remaining 14%, field crops such as grassland and fodder and outdoor vegetables, account for over half. All measurements are expressed as tonnes of active ingredient (ai) applied, although the area treated is often also used as a measurement of usage.

Table 1. Distribution of pesticide usage by weight - data for 1993 by crop type

Crop Type	Pesticide Usage (tonnes ai)
Total	32,118
arable crops	27,746
grassland & fodder	1,744
outdoor vegetables	883
top fruit	352
glasshouse crops	296
mushrooms	277
soft fruit	247
hops	236
outdoor bulbs	229
hardy nursery stock	108

Within the figures given in table 1 and leaving aside sulphuric acid, which is applied as a herbicide and crop desiccant, pesticides usage measured by tonnes of active ingredient applied can be ranked by type as herbicides > fungicides > growth regulators > insecticides as shown in table 2.

Table 2. Distribution of pesticide usage by weight - data for 1993 by pesticide type

Pesticide Type	Usage (tonnes ai)
sulphuric acid	10,167
'other' herbicides	10,668
fungicides	6,862
growth regulators	2,665
insecticides	1,139
'other' pesticides	635

Such data, and the more detailed information presented annually in the published reports on pesticide usage[2] which provide information on individual active ingredients, are invaluable tools for the regulators and their analysts to target resource on the analysis of what has been used. However, data for a single year do not show the changing patterns in usage brought about by the availability of new compounds, and regulatory and environmental pressures.

The availability of new chemicals has had some dramatic effects on usage, and hence residues, in the United Kingdom. An example of this change is well illustrated by the use of a 'pair' of herbicides, fungicides and insecticides and comparing what was used in 1983 and 1993. Less than 1% of the weight of herbicide applied to oilseed rape in 1983 is applied now. Weed control is achieved using new chemicals with higher biological activity but applied at reduced levels. The same phenomenon can be observed in the use of fungicides on apples and field insecticides on wheat.

Table 3. Some examples of product changes resulting in major reductions in weight applied

1983		*1993*
TCA		fluazifop-P-butyl
	oilseed rape	
133,718 ha		153,191 ha
1,942,211 kg		15,040 kg
1.45 kg/ha		0.09 kg/ha
captan		myclobutanil
	apples	
46,669 ha		59,616 ha
127,870 kg		2,772 kg
2.75 kg/ha		0.05 kg/ha
D-S-M		cypermethrin
	wheat	
193,325 ha		825,050 ha
34,798 kg		19,599 kg
0.18 kg/ha		0.024 kg/ha

All the data given so far refer to 1993 or to selected pesticides classes over a ten year period. Can the sustained policy of minimising pesticide usage in agriculture, commensurate with the safe and efficient production of food, be measured over a range of crops and pesticides? Data for insecticide usage on cereal crops in 1982, 1988, 1990 and 1993[3] are shown in figure 1. The graph demonstrates the changes that have taken place in the 13 year period. The use of organochlorine compounds has declined to be partly replaced, initially by organophosphorus pesticides and most recently by the use of synthetic pyrethroids.

Figure 1. Comparison of insecticide usage on cereals in Great Britain 1982 - 1992, amount used (tonnes)

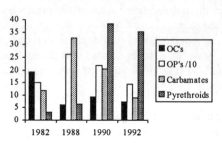

A refinement in plotting long-term trends is the use of the 'average environmental loading'. This is a measure of long-term pesticide usage which takes into account the rate at which a pesticide formulation is applied as a proportion of the manufacturers' recommended rate. As such, the measure is independent of changes in the product spectrum.

average environmental loading = $\underline{\Sigma((farmer's\ rate\ /\ recommended\ rate)\ x\ f.area\ treated)}$
$$\Sigma\ (f.area\ treated)$$
{where f.area is the formulation area in kg/ha.}

Data for agricultural pesticide use in England and Wales (figure 2) confirms that there has been a steady reduction in the average environmental loading of pesticides, a trend which continues with the preliminary 1994 figure.

Figure 2. Relative environmental loading on all arable crops (except sugar beet)

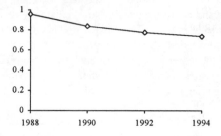

These data provide the regulator and analysts with information upon which to base residue monitoring strategies in foods and in the environment. However, changes in usage patterns and the appearance of new chemicals are not the only challenge facing the analyst.

2.2 Changes in the regulation

Regulations governing the maximum residues levels (MRL) of pesticides permitted on foodstuffs produced to Good Agricultural Practice (GAP) and in the environment have increasingly become a matter of international interest. United Kingdom regulations, Statutory Instrument 1378 (1988)[4], recently superseded by Statutory Instrument 1985 (1994)[5] are relevant to those analysing for residues in that they indirectly set which residues are determined and the levels at which analytes should be sought. However, such legislation is inevitably influenced, if not dictated, by European Directives such as 94/29EC[6] concerning residues in cereals and foods of animal origins and 94/30EC[7] concerning residues in fresh produce, and those Directives which preceded them. Other international bodies such as the Codex Alimentarius Commission, a joint UN Food & Agriculture Organisation / World Health Organisation body, also play a key role in setting MRL's.

The analyst is having to respond to the changing needs of legislation in two ways. The number of pesticides residues, their breakdown products and metabolites which have to be sought continues to increase. The European Commission has an on-going programme reviewing MRL's and is continually adding more through revised Directives. The UK, whilst not necessarily immediately mirroring the Commission, is obliged to ultimately implement European Directives in UK legislation. For the laboratory it means more residues to analyse for in more foodstuffs. For example SI 1378 had 11 pages and covered about 50 residues in a variety of 40 or so foods. By comparison SI 1985 taking into account European legislation, has over 40 pages and regulates the permissible levels of some 90 residues in about 145 foodstuffs.

However, in addition to the range of residues to be sought, legislation is also calling upon the analyst to seek residues at lower levels. SI 1985 is peppered with asterisks indicating that the MRL quoted is "at or about the limit of determination". Notably, with some international regulations, this tendency can lead to an anomaly where regulations cannot be 'enforced' since suitable analytical methods do not exist to generate <u>confirmed</u> data for the parent and any products at the necessary levels. Whilst increasing numbers of MRL's include metabolites and/or breakdown products, satisfactory methods for the parent molecule may not be suitable for determining the range of breakdown products at the MRL. This can be compounded by the lack of suitable analytical standards for breakdown products and hence the inability to calculate the recoveries for a true estimate of the levels of pesticide present. The MRL for the non-systemic fungicide vinclozolin includes its 3,5-dichloro aniline breakdown products. Not only are suitable standards not available for all of the potential products but 3,5-dichloro aniline can arise from the breakdown of other pesticides including iprodione and procymidone. This makes it almost impossible to provide reliable data in support of MRL monitoring in this instance.

The combination of more residues to be sought, at lower levels and together with a greater number of metabolites and/or breakdown products is a challenge to the capabilities of the analyst. Coupled with the need for unequivocal confirmation of data and, increasingly, an eye to making the best use of necessarily finite, if not limited resources, the task truly becomes a challenge.

3. CONCLUSION - FACING THE CHALLENGES

There is a battery of sophisticated techniques which the analyst may use to meet the challenges of changes in agricultural practices and legislation. Foremost amongst these are the hyphenated techniques such as gas chromatography - mass spectrometry (GC-MS), liquid chromatography - mass spectrometry (LC-MS) and capillary electrophoresis - mass spectrometry. These separation techniques, each applicable to groups of pesticides with differing physical properties, coupled to mass spectrometry for both unequivocal identification of analytes and broad detector specificity offer powerful tools for residues analysis. The properties of some analytes continue to demand individual analysis. However a costly but efficient option, GC-MS (ion trap) with appropriate use of selected ion monitoring, can determine over 100 pesticides in a sample as a single analysis.

The ability of the analyst can lag behind changes in practice. For example, the triazine herbicides are increasingly being replaced by glyphosate. However, robust methods for the analysis of glyphosate in foods and environmental samples are only now becoming available. Notwithstanding this, it must be hoped that policy and regulation can take into account the practical implications of decisions on the analyst who may have to provide data to support monitoring and enforcement actions. Likewise, the advances in analytical science may be important when re-evaluating existing legislation. There is a great debate in the United States over the impact of the "Delaney Clause" on pesticide regulation. Briefly, the clause states that no trace of a chemical suspected of having carcinogenic properties is permitted in processed foods. The clause was drafted in 1958. Since then the level of what analysts deem a 'trace' has been significantly reduced. Opponents of the clause cite this, together with other factors, as the basis for replacing the clause with something more similar to the view that "regulation should focus on those instances where there is clear evidence of potential risk". Here the skills of the analyst are having a direct effect on the political debate.

It is vital that the efforts of the analyst are focused on meeting the needs of the regulator for unequivocal data in support of monitoring and enforcement exercises. However, it is equally vital that the regulators, on an increasingly international scale, seek the views of the analyst before enshrining anomalies and impossibilities in regulations.

References

1. C Tomlin, ed., 'The Pesticide Manual', 10th Edition, Crop Protection Publications, Farnham, 1994

2. R P Davis, D G Garthwaite and M R Thomas, (1990) Pesticide Usage Survey Report 85 - Arable Crops 1990, MAFF London

3. R P Davis, M R Thomas, D G Garthwaite and H M Bowen, (1992) Pesticide Usage Survey Report 108 - Arable Crops 1992, MAFF London

4. Anon., Statutory Instrument 1378 - The Pesticides (Maximum Residue Levels in Food) Regulations 1988, HMSO, London

5. Anon., Statutory Instrument 1985 - The Pesticides (Maximum Residue Levels in Crops, Food and Feeding Stuffs) Regulations 1994, HMSO, London

6. Anon., Council Directive 94/29/EC - amending the Annexes to Directives 86/362/EEC and 86/363/EEC on the fixing of maximum levels for pesticide residues in and on cereals and foodstuffs of animal origin respectively, *OJ No L 189 67 - 69, 1994*, Brussels

7. Anon., Council Directive 94/30/EC - amending Annexe II to Directive 90/642/EEC relating to the fixing of maximum levels for pesticide residues in and on certain products of plant origin, including fruit and vegetables and providing for the establishment of a list of maximum levels, *OJ No L 189, 70 - 83, 1994*, Brussels

Non-agricultural Uses of Herbicides: Triazines to Glyphosate

R. P. Garnett

MONSANTO PLC, THAMES TOWER, BURLEYS WAY, LEICESTER LE1 3TP, UK

1 INTRODUCTION

Herbicides have been used to control weeds in "non-agricultural" areas for around 50 years. Non-agricultural uses cover very diverse situations from forestry to parks and gardens, from waterways to roads and railways, and from pavements to industrial sites. Not only are the uses varied, but also the users, many of whom have little knowledge of plants, weed control or the use of herbicides.

Perhaps the greatest volume of herbicide is used to maintain man-made surfaces such as railway ballast, road edges, pavements and channels, gravelled areas etc. These are now officially called "land not intended to bear vegetation" by the Pesticides Safety Directorate. They are often referred to as "hard surfaces", and can be divided into porous or non-porous surfaces.

Table 1 *Survey of non-agricultural pesticide use in 1990[1]*

Type of User	% Total Volume	Active Ingredients	% Total Volume
power & industry	32	triazines	39
local authority	31	diuron	7
transport	19	other residuals	10
forestry	9	glyphosate	5
water companies	5	other non-residuals	39
golf & leisure	4		

2 NON-AGRICULTURAL HERBICIDES

The first significant herbicides for non-agricultural weed control were based on sodium chlorate, which was the most widely used non-selective herbicide[2,3] up to the mid or late 1960's. It was used at very high rates, 56-448 kg ai/ha, but even higher rates of boric acid, another herbicide, were applied, 1180-5020 kg ai/ha. New chemistry began to be introduced in the late 1950's (Table 2), which was applied at lower rates of active ingredient to achieve the season long weed control which was required. Most of these herbicides were applied in the range of 3.5-35 kg ai/ha, the higher rates being used for initial site clearance, and the lower rates for maintaining bare ground. By the mid 1960's most of today's armoury of herbicides for man-made surfaces had been introduced.

These herbicides were mainly designed for agricultural use, but were later developed for non-agricultural uses. Their expense compared to sodium chlorate precluded significant use until the early 1970's. Then simazine and atrazine came off patent, and alone, or mixed with amitrole, began to dominate usage. These combinations, and variations with the other herbicides in Table 2 and hormone weedkillers, proved extremely effective. The reliance on those products remained until the late 1980's, but then came the realisation that triazine herbicides were regularly being found in drinking water at significant levels.

3 THE PROBLEM

Regular monitoring for pollutants, including pesticides, in water has been carried out for a number of years but began in earnest with the implementation of the Water Supply (Water Quality) Regulations, 1989, which incorporated the regulations of the EC Drinking Water Directive (80/778/EEC). The standards for pesticides are well known: 0.1 μg/l for an individual pesticide, and 0.5 μg/l for the total pesticide content.

Three non-agricultural herbicides, atrazine, simazine and diuron, were soon identified as a major source of exceedences of the standards[5], although they account for only 2-5[6,7]% of the total volume of herbicide active ingredient applied each year in the UK. There are several factors which are likely to have contributed to this:

Table 2 *Major non-agricultural herbicides*[3,4]

	introduced in:	persistence:
simazine	1956	season or longer
monuron	1957	" " "
dalapon	1957	3-4 months
atrazine	1958	season or longer
amitrole	1960	1-2 months
diuron	1960	season or longer
paraquat	1962	none
dichlobenil	1965	season
bromacil	1965	season or longer
picloram	1967	season or longer
glyphosate	1976	none
imazypyr	1985	several months

Table 3   *Sorption and persistence classification for selected herbicides[6] and the chance of reaching water[8]*

Pesticide	Adsorption (Koc cc/g)	Mobility class	Half Life (days)	Persistence class
likely to reach water				
dicamba	2	Very mobile	14	Slightly
mecoprop	20	Mobile	21	Moderately
2,4-D	20	Mobile	10	Slightly
atrazine	100	Moderately mobile	60	Moderately
simazine	130	Moderately mobile	60	Moderately
amitrole	200	Moderately mobile	14	Slightly
diuron	480	Moderately mobile	90	Very
unlikely to reach water				
glyphosate	10,000	Non-mobile	30	Moderately
paraquat	100,000	Non-mobile	500	Very

- The characteristics of residual herbicides, which are designed to be persistent and resist degradation to achieve long term weed control (Table 3).
- Usage on surfaces susceptible to wash off directly into surface water systems (e.g. from pavements, channels and drains) or to leaching into ground water (particularly shallow or rapidly recharging ground water). Hard surfaces reduce the opportunities for microbial degradation, compared to soil. The mechanisms for movement of herbicides from man-made surfaces to water are poorly understood, but vulnerability studies, such as CatchIS[9], and other recent initiatives should greatly improve knowledge.
- Application by poorly trained spray operators, guided by poorly written specifications, can easily lead to overdosing, inaccurate spraying and spray drift. Operators sprayed directly over drains or culverts, and it was common to spray in the rain "because it activates the weedkiller" even though this gives the highest risk of wash-off. The standard of application is improving, but varies greatly.
- Specifications for weed control often required "total" weed control for the whole year; consequently not only were the label recommended doses high, but overdosing was encouraged to minimise the need to re-spray.
- Incorrect disposal of excess spray, sprayer washings or contaminated packaging, or spillage of concentrate herbicide while preparing the spray mixture.

Over 2.5 million determinations were made by water supply companies in 1993, and 1.4% showed exceedences. Pesticides accounted for 42% of all those analyses, with 2.4% showing exceedences. Pesticides were the cause of 73% of all exceedences[10], but there were distinct regional differences (Table 4). Where water is abstracted from lowland rivers, the level of contraventions was greater than where abstractions are primarily from upland catchments or ground water. For example, pesticides accounted for 94% of all exceedences in the Thames catchment, where pesticides accounted for 68% of all analyses undertaken. Analyses for pesticides in Thames alone accounted for 23% of all

Table 4   *Exceedences of standards for drinking water, 1993*[10]

Supplier	All determinations % exceedence	Pesticide determination % exceedence	% all exceedences due to pesticides	% pesticide exceedence due to non-ag. herbicides
England & Wales	1.4	2.4	73.4	-
Anglia	0.7	0.2	9.5	66.3
North West	0.5	0.2	3.1	63.6
Northumbrian	0.4	0.1	4.6	0.0
Severn Trent	0.3	0.2	32.4	22.0
Southern	0.5	0.6	52.1	66.7
South West	0.5	0.0	1.3	50.0
Thames	5.3	7.3	93.9	67.2
Welsh	0.6	0.0	2.4	28.5
Wessex	0.3	0.4	15.3	64.5
Yorkshire	0.6	0.2	19.2	3.52

pesticide determinations in England and Wales. Pesticides also accounted for a high level of exceedences in Southern and Severn Trent, but were less significant in other areas.

Non-agricultural pesticides (atrazine, simazine and diuron) accounted for ≥ 50% of pesticide exceedences in 6 of the 10 major supply regions (Table 4). One region, the Thames catchment, accounts for about half of all non-crop herbicide determinations carried out in England and Wales, 98% of all non-crop herbicide exceedences, and about three quarters of non-compliant zones.

The exceedences shown in drinking water are mirrored by analyses of surface freshwater and ground waters by the National Rivers Authority (Table 5).

4 ACTIONS TAKEN

Water supply companies are required by the Water Industry Act, 1991 "to supply only water which is wholesome at the time of supply". Water which exceeds the standards is not "wholesome" so the supply companies are obliged to take action.

The two water supply companies with the greatest number of exceedences due to non-agricultural herbicides have both undertaken large scale programmes to influence herbicide use within their catchments. The objective was to reduce herbicide contamination by introducing better weed control practice as part of a plan for a cost effective, integrated strategy to meet the drinking water standards[12].

Non-agricultural uses were the major focus, not only because of the high proportion of exceedences but also because of the well defined, relatively small number of users.

· Firstly, a pesticide usage survey was essential so that a strategy could be developed, and a focussed monitoring scheme be established, as required by the regulations.

· Secondly, high profile catchment protection measures were initiated, and both organisations established direct contact with a large number of users, and worked with manufacturers to improve awareness of the problems and to provide guidance.

Table 5   *Pesticide exceedences reported by the National Rivers Authority.*
              Source: NRA

	1992		1993	
	Total number analyses	% samples > 0.1 μg/l	Total number analyses	% samples > 0.1 μg/l
SURFACE FRESHWATER				
atrazine	3965	17	4100	14
simazine	4065	13	4094	9
diuron	693	14	1598	18
GROUNDWATERS				
atrazine	531	9	603	11
simazine	523	1	603	2
diuron	104	0	129	5

•   Thirdly, research on catchment vulnerability was initiated to identify factors which affect the process of pesticides reaching water, to allow the identification of vulnerable areas, and to predict pollution risks.

•   Finally, water treatment is necessary to ensure compliance with the standard. Catchment protection takes time to implement to achieve consistent results and it is unlikely to achieve complete protection. New treatment processes have been installed, and while there are other water quality benefits, the scale, size and cost of plant required is often dictated by pesticide contamination levels.

These programmes began in 1990 when Thames Water wrote to all local authorities and British Rail regions in their catchment asking them to review their use of herbicides and to use alternatives to residual herbicides. Responses emphasised the low level of technical expertise available, the uncertainty over which herbicides were used, and that there was considerable scope for reducing herbicide use[11]. Severn Trent initiated a parallel programme in 1992 leading to "Spraysafe: a charter for non-agricultural use of herbicides"[12]. This encouraged a rational, integrated approach to weed control, a reduction in the use of residuals and the correct use of herbicides.

The government also took action. The sale and advertising by manufacturers of simazine and atrazine for non-agricultural uses was revoked in 1992, and usage was banned from September 1993. In addition, the Department of the Environment published a guide to managing vegetation in non-agricultural situations with regard to environmental and water protection[13], together with a complementary guide for users[14].

Manufacturers became closely involved in the education programme. Monsanto initiated an intensive promotional campaign on weed control in 1991, to ensure that a practical alternative to residual herbicides was available. This was based on glyphosate, a leaf acting herbicide, which had been identified as a suitable alternative to residuals because it was considered "unlikely to reach water"[8]. Glyphosate has several features which allowed water supply companies and other experts to reach this conclusion:

-      it is rapidly and tightly adsorbed to soil or sediment[15].

-      glyphosate is readily degraded by microorganisms in biologically active systems such as soil and water, with a typical half-life of 10-60 days depending on conditions[16].

-    glyphosate is one of the few active ingredients approved for weed control in or near water due to its favourable aquatic toxicology profile[15]; this profile has now been further improved by the introduction of a new formulation with a more benign surfactant[17].

Rhône Poulenc, major manufacturers of diuron, initiated a campaign to manage the use of diuron as an alternative residual herbicide. It had become obvious that atrazine was being directly replaced with diuron, and that there was a new risk of water contamination. The programme was directed to reduce the dose of diuron applied, to avoid application to drains and channels, and to treat only in joints and cracks rather than over slabs and tar macadam[18].

4 RESULTS
4.1 New weed control practice

The combined effect of these efforts was a revolution in weed control operations and herbicide use in non-agricultural areas. The promotion of reduced usage or replacement of residuals, and the revocation of the use of triazines, has caused problems for most of those involved in weed control. However, there have been important benefits:
-    attention was focused on weed control, a "cinderella" subject in a world dominated by engineers and others with little biological training; clients and specifiers for weed control programmes are now making efforts to raise their level of understanding;
-    weed control objectives have been reviewed: "why control weeds" and "how much weed growth is acceptable" are obvious questions which have often remained unanswered;
-    weed control programmes have been reassessed.

A further result was a plethora of "experts" offering advice - manufacturers, contractors, consultants, etc. Consultations between the various parties developed new guidelines which were published by the Department of the Environment[12,13]. These advocated an integrated approach to weed control, using a range of techniques to minimise the risk of contaminating water sources with herbicides.

A variety of non-chemical methods is available. Many of these have long been standard practice, such as the use of roadsweepers, but there are other opportunities which can be used as the basis of a weed control programme[19,20]. For example, optimum landscape design can minimise weed problems at source, and good maintenance reduces subsequent problems. Even pedestrians can be utilised as excellent weed control agents by constantly walking over surfaces.

However, herbicides usually complement these methods to ensure high levels of weed control. Relatively few herbicides are approved for non-agricultural weed control, and even fewer can be used on pavements and in channels[21]. In practice these are: glyphosate, amitrole and diuron. In areas without trees or other desirable plants, bromacil and imazapyr can be used. There is a recent tendency to introduce hormone weedkillers such as MCPA into spray mixes, although these are not clearly approved for use on hard surfaces.

The core of most weed control programmes is now glyphosate. The use of a leaf acting product without a residual herbicide has enforced a new approach to weed control. Obviously weeds must be present at the time of control, unlike when residuals are used, and weeds can germinate immediately after treatment to reinfest the area. Consequently, a two or three spray programme is used. The first application is usually in spring, and the last in early autumn. Sometimes another is needed in summer, depending on the level of weed growth and the maintenance objectives. Multiple applications need not be more expensive than old practice if they are part of an integrated approach to weed control.

Financial gains are possible, and weed control may be better.For example, one local authority had budgeted for treating the whole of its pavement area, but then undertook its first weed survey. This showed that only about 30% of the area warranted treatment, because some was no longer paved, some kerbs had recently been re-set in concrete, and regularly swept parts of the town centre had few weeds. In addition, glyphosate had controlled deep rooted weeds which had become established during previous seasons.

Although many local authorities and contractors have found the changes in weed control practice to be difficult to manage, some have illustrated that the change to glyphosate achieves outstanding weed control and helps their overall weed control programme. The key is to understand weed growth, to understand the herbicide and its use, to monitor the results carefully, and to be flexible in the weeding programme. They have enshrined these principles in their specifications for the weed control operations.

New practices continue to develop. Residuals used on railways have been clearly shown to reach water (*e.g.* atrazine in Switzerland[22]), and Railtrack are working with the water industry to minimise their impact on water quality. The standard rail treatment is now glyphosate with diuron, but glyphosate alone is being tested prior to more widespread use in vulnerable areas. One of the difficulties is the practicality of arranging for the spray train to apply multiple treatments, when one spray per year is normal. However, this may not be necessary on all lines.

4.2 Reduction of exceedences due to herbicides

The quantity of triazines used fell from 39% of the volume of active ingredient used in non-agricultural areas in 1991[1] to 9% in 1993[18]. In contrast, diuron increased from 7% to 24% over the same period. Glyphosate is the only product used for weed control on hard surfaces such as roads and pavements by an increasing number of local authorities.

These changes resulting from the educational and promotional campaigns have led to significant declines in residues found in water. This is clearly indicated by data supplied by Severn Trent[12] and Thames Water[23], and in the Drinking Water Inspectorate Annual Reports[5,10].

The national total of exceedences for triazines as a proportion of the number of determinations was fairly constant from 1990 to 1992, but fell in 1993[18]. Diuron, however, increased from 1990 to 1992 but also fell in 1993. Analyses by the NRA showed a similar fall in the proportion of exceedences for triazines in surface waters but the figure for diuron increased from 14 to 18% (Table 5).

In the regions where extensive campaigns against residuals were carried out, there were significant reductions in contamination. The yearly exceedences in raw waters in the Severn Trent region fell sharply for both triazines and diuron, from 10-26% of samples for individual herbicides in 1990 to 0.4-2.3% of samples in 1994 (Figure 1)[12]. Meanwhile, the problem of agricultural herbicides such as isoproturon has remained at a high level.

In the Thames catchment area peak levels of atrazine and simazine also fell sharply between 1990 and 1994. The data in Figures 2 and 3 are from drinking water taken from a surface water treatment works where no advanced water treatment has been operational[19]. There is an annual peak contamination in mid summer, after the main period of use. Typically there seems to be a 30-40 day time lag from application to appearance in water.

Since the Thames Water catchment protection programme began in 1990 there has been an annual fall in residues of the triazines to below the 0.1 μg/l standard (Figure 2). The small decline between 1992 and 1993 is thought to reflect efforts to use up stocks of

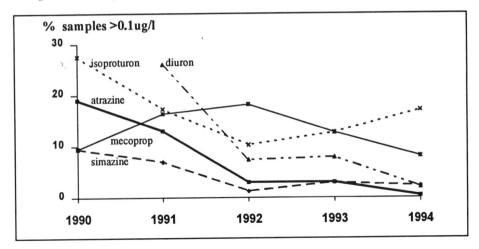

Figure 1 *Pesticide levels in raw water: Severn Trent*

product before the ban was implemented. Some illegal use in non-agricultural areas is suspected, and there are legal uses in maize and in garden products.

For diuron, there was a gradual increase in residues from 1990 to 1993 reflecting its use to replace the triazines (Figure 3). In 1994 there was a sharp fall to below the 0.1 μg/l standard. Although this remains to be confirmed by analyses in subsequent years, this is felt to reflect the impact of the campaign to improve the use of diuron, and the move to glyphosate alone by many local authorities.

In fact in 1994 there were few exceedences by non-agricultural herbicides. The two triazines and diuron together totalled 84% of exceedences in Thames in 1990, but only 23% in 1994[23]. In contrast, the agricultural herbicide isoproturon increased from 10% to 68% of exceedences during the same period.

The detection of diuron in ground water for the first time (Table 4) is worrying. It is speculative as to whether this is the result of rapid transfer to shallow ground waters after the increase in use of diuron over the past four years.

In 1993 a survey identified that the main use of diuron was on porous (e.g. gravel) surfaces rather than non-porous surfaces such pavements[18]. There seemed to be a link between the occurrence of diuron residues in water, and the few locations in which it had been used on pavements, in channels and over drains. The survey concluded that direct introduction of diuron at high rates into drains and channels was the main route for water contamination. Applications to porous surfaces or incorrect disposal were not significant causal factors.

5 CONCLUSIONS

Water monitoring has shown that non-agricultural herbicides have great potential to contaminate water catchments, particularly lowland rivers. However, the recent falls in the number of exceedences attributed to these herbicides suggest that appropriate catchment protection schemes can be very successful.

The choice of herbicides approved by the Government for use on hard surfaces is very limited, particularly on pavements where trees are growing. There is always the

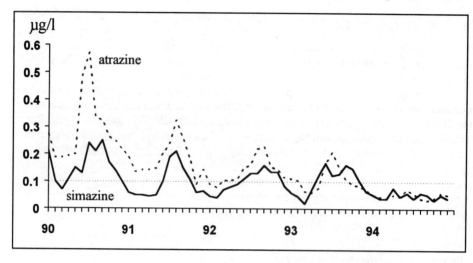

Figure 2 *Concentrations of atrazine and simazine in drinking water supplies from a water treatment works serving West London*

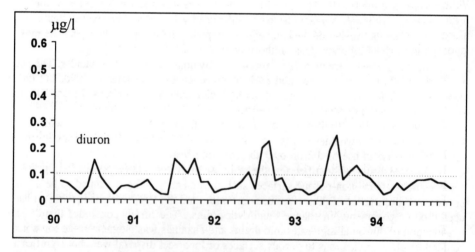

Figure 3 *Concentrations of diuron in drinking water supplies from a treatment works serving West London*

danger of excessive use of one residual causing contamination, as occurred with diuron as the use of triazines was discouraged and finally banned. New active ingredients are being tested for this market which may change the situation, but the experience with existing residuals has increased the stringency of the approvals process. The concept of "designer" herbicides has been introduced to help solve the problem[24].

Improved practice through proper training of specifiers and users is essential to minimise the risk of future problems. In 1995 training schemes are being introduced by the British Agrochemicals Standards Inspection Scheme (BASIS) and the Agricultural Training Board to complement the efforts of the water industry and the manufacturers. Good specifications for weed control are basic to success, and several organisations have produced "guideline specifications" to help specifiers ensure that weed control operations are carried out to high standards. Unfortunately, a good specification cannot achieve this without appropriate monitoring of the operations - quality assurance. For this to work, all involved must talk to each other - a surprisingly difficult objective to meet. Fortunately, the change to a completely different type of herbicide, glyphosate, has necessitated a thorough revision of weed control practices which has ensured at least some communication between different organisations.

Research over the next few years will expand knowledge on the mechanisms of herbicide transfer from hard surfaces to water, and will allow improved risk assessment through catchment vulnerability models. This information, and advice on weed control practice and herbicide usage must be disseminated to those involved in weed control. While the exercises described in this paper have been largely successful, there still seem to be those who are not aware of the legislation and the problems. Dissemination of information must be efficient so that everyone is aware of the pitfalls, changes and future options, and that they have time to respond by modifying their practice.

References

1. "The Use of Herbicides in Non-Agricultural Situations in England and Wales", Department of the Environment, 1991.
2. British Weed Control Council, "Weed Control Handbook", Blackwell Scientific, London, 1958, p.13.
3. E. K. Woodford and S. A. Evans (editors), "Weed Control Handbook", 3rd edition, Blackwell Scientific, London, 1963, p.14.
4. R. J. Hance and K. Holly (editors), "Weed Control Handbook: Principles" 8th edition, Blackwell Scientific, London, Chapter 3, p.75.
5. Drinking Water Inspectorate, "Drinking Water 1990", HMSO London, 1991, pp.194.
6. A. D. Carter, "Aspects of Applied Biology 29, 1992: Vegetation Management in Forestry Amenity and Conservation Areas", Association of Applied Biologists, 1992, p 17.
7. British Agrochemicals Association "Annual Report 1994", p.28.
8. Department of the Environment/Welsh Office, "Guidance on Safeguarding the Quality of Public Water Supplies", HMSO London, 1992, p. 99.
9. J. M. Hollis, C. A. Keay, S. H. Hallet, J. W. Gibbons and A. C. Court, "Monograph No. 62, Pesticide Movement to Water", British Crop Protection Council, 1995, p. 345.
10. Drinking Water Inspectorate, "Drinking Water 1993", HMSO London, 1994, pp.250.

11. S. L. White and D. C. Pinkstone, D.C., "Brighton Crop Protection Conference - Weeds - 1993", British Crop Protection Council, 1993, p. 363.
12. A. C. Court, R. A. Breach and M. A. Porter, "Monograph No. 62, Pesticide Movement to Water", British Crop Protection Council, 1995, p. 381.
13. Department of the Environment, "Weed control and environmental protection", HMSO London, p. 46.
14. Department of the Environment, "Guidance for control of weeds on non-agricultural land", p. 28.
15. L. Torstensson, "The Herbicide Glyphosate", E. Grossbard, D. Atkinson (editors), Butterworths, London, 1985, p. 137.
16. T. E. Tooby, "The Herbicide Glyphosate", E. Grossbard, D. Atkinson (editors), Butterworths, London, 1985, p. 206.
17. R. P. Garnett, "Proceedings of the Symposium on Aquatic Weeds, Dublin, 1994", European Weed Research Society (in press).
18. A. B. Davies, R. Joice, J. A. Banks and R. L. Jones, "Monograph No. 62, 1995, "Pesticide Movement to Water", British Crop Protection Council, 1995, p. 311.
19. R. P. Garnett, "National Turfgrass Council Workshop Report, No. 24: Amenity Management Double Bill", National Turfgrass Council, 1994, p. 87.
20. J. Hitchmaugh, "National Turfgrass Council Workshop Report, No. 24: Amenity Management Double Bill", National Turfgrass Council, 1994, p. 112.
21. R. P. Garnett, "National Turfgrass Council Workshop Report: Correct Application, not Complete Abdication", 1995 (in print).
22. A. Seiler and F. Muhlebach, "Monograph No. 62, Pesticide Movement to Water", British Crop Protection Council, 1995, p. 389.
23. S. L. White and D. C. Pinkstone, "Monograph No. 62, Pesticide Movement to Water", British Crop protection Council, 1995, p. 263.
24. R. Haq and J. M. Perkins, "Brighton Crop Protection Conference - Weeds - 1993", British Crop Protection Council, 1993, p.417.

Synthetic Pyrethroids Success Story

R. M. Perrin

ZENECA AGROCHEMICALS, FERNHURST, HASLEMERE, SURREY GU27 3JE, UK

1 INTRODUCTION

The synthetic pyrethroid insecticides can almost be described as too successful, such is their world-wide reputation for cost-effective, broad-spectrum pest control. Excellent news for users and for company profits, you might say, but in fact over-use in some markets threatens their longer-term viability and has on occasions led to the abandonment of other chemical and biological protection measures. Instead of embracing the rationale of integrated pest management (IPM) on which sustainable agriculture and environmental stability ultimately depend, many growers have resorted to repetitive application of pyrethroids as a single weapon against insect pests. Cotton farmers in particular revolutionised their crop protection around the pyrethroids, the emergency registrations of permethrin and fenvalerate in 1977 being heralded in Texas "in the same light as their forefathers greeted the 7th Cavalry!"[1]

"No insecticide is perfect, but the pyrethroids do come close."[2] Few synthetic pesticides offer economic and social benefits without potential risks to users, consumers or the environment, but pyrethroids have, since their introduction in the mid-1970's, proved a powerful insect control tool with little evidence of the acute hazards or longer term impact associated with most earlier chemical groups. Consequently, great commercial success has been achieved in two decades of use, with agricultural sales of around $1.4 billion still growing at 1.5% per annum[3], plus substantial sales in public and animal health, industrial, amenity and household outlets, as well as in grain and food stores.

This paper highlights the reasons behind the pyrethroid success story to date, predicts likely trends in the future, and discusses some of the technical and political hurdles which must be faced in developed and developing countries.

2 WHAT IS A PYRETHROID ?

The term "pyrethroid" is generally reserved for the so-called photostable synthetic derivatives of natural pyrethrins, made and tested in several laboratories but most notably by Elliott and co-workers at Rothamsted Experimental Station[4] and by Sumitomo in Japan. A few earlier, unstable examples, including allethrin, tetramethrin and resmethrin, exhibited excellent insecticidal activity and found use in household, stored-product and veterinary products. Sumitomo synthesised and patented the first light-stable pyrethroid of agricultural interest, fenpropathrin, in 1971, although it was not developed and sold until 1980. The first compound to be commercialised, permethrin, was announced in 1973[5], followed by a succession of more potent analogues, still being patented today[6], from the public and private sector. In less than a decade of commercial development, pyrethroids became one of the top four classes of synthetic insecticides.

All pyrethroids are structurally similar (Figure 1), derived from the constituent esters of pyrethrum, and share a common mode of biological action, namely disruption of voltage-gated sodium channels on nerve axons. There is a diversity of insecticidal products with respect to the optical and geometric configurations of their acidic and alcoholic components, number and ratio of isomers, and physico-chemical properties. A non-volatile, moderately persistent molecule like cypermethrin has very different uses to a volatile, non-persistent molecule like tefluthrin. Not all compounds labelled as pyrethroids nowadays are, in fact, esters, since several ethers have been discovered with similar insecticidal properties but often less intrinsic toxicity to aquatic organisms.

3 COMMERCIAL SUCCESS

Pyrethroids are blessed with a very broad-spectrum of arthropod activity, including most of the major pest groups, at exceptionally low dosages, coupled with relative lack of hazard in practice to users and the environment. This has allowed registration and sales in a wide range of markets, as illustrated in Table 1. Comparing the typical use rates and toxicological profiles of pyrethroids, carbamates and organo-phosphates (Table 2), it is hardly surprising that pyrethroids are the first choice in so many situations. Dosages per hectare are especially low where a single active cis isomer or enantiomer pair are employed Deltamethrin,(S)-α-cyano-3-phenoxybenzyl(1R)-cis-3-(2,2-dibromovinyl)-2,2-dimethylcyclopropanecarboxylate, is a well-known example of resolution to the single biologically-active isomer. Five grams of deltamethrin or lambda-cyhalothrin can, for example, protect the same area of cereals from aphid damage as half to one kilogram of an organophosphate, and fifteen kilograms can treat as many houses for mosquito control as one tonne of DDT. The main competitive advantages of pyrethroids, promoted heavily by manufacturers and suppliers, are listed in Table 3.

Table 1   *Typical Markets for Pyrethroid Insecticides*

Use	Examples of Pests Controlled
Cotton	Bollworms, Armyworms
Vegetables	Armyworms, Pinworms, Fruit borers
Cereals	Aphids
Hospitals, Restaurants	Cockroaches, ants
Dwelling Places	Mosquitoes
Bed nets	Mosquitoes
Building foundations	Termites
Cattle and sheep dips	Ticks, lice
Carpets	Fleas
Lure and kill traps	Bollworms, cockroaches
Aerosols	Flies, mosquitoes
Grain stores	Grain beetles, flour moths
Timber	Wood-boring beetles

Table 2   *Comparative Toxicology and Use Rates for Pyrethroids, OP's and Carbamates*

Chemical Group	Typical Range for Technical Material		Typical Range of Field Use Rates (g ai/ha)
	Acute oral LD_{50} (rat) (mg/kg)	Acute dermal LD_{50} (rat) (mg/kg)	
Pyrethroids	100-5000	1000-5000	5-100
OP's	10-500	50-3000	250-1500
Carbamates	20-100	1000-5000	125-1000

Figure 1   Some Examples of Pyrethroid Insecticides

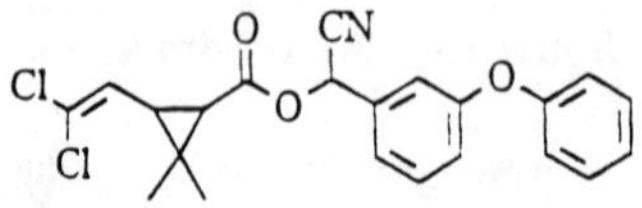

Cypermethrin

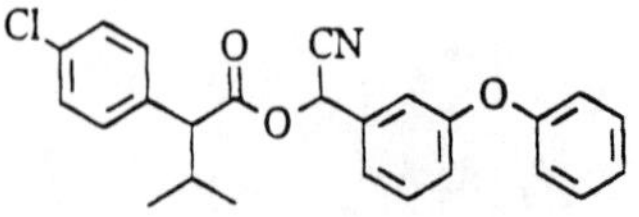

Fenvalerate

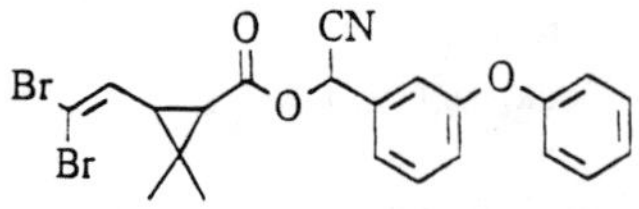

Deltamethrin

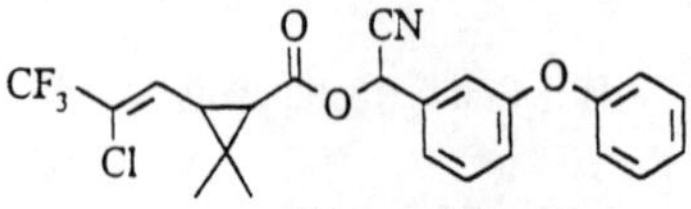

Lambda-cyhalothrin

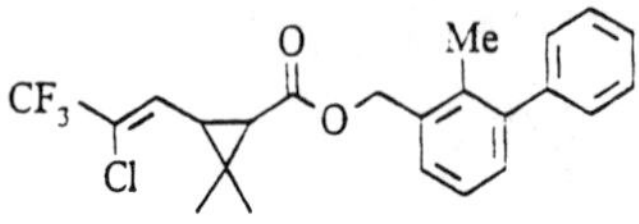

Bifenthrin

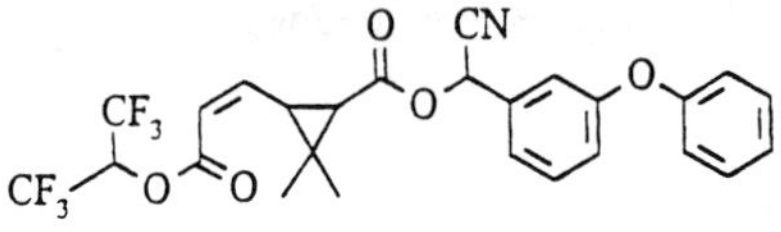

Acrinathrin

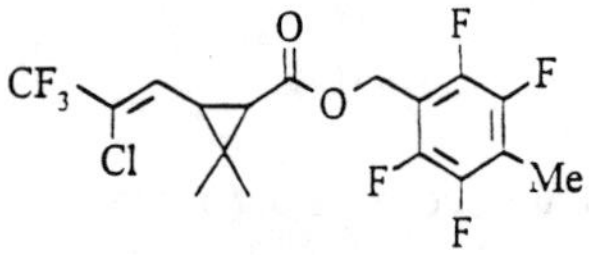

Tefluthrin

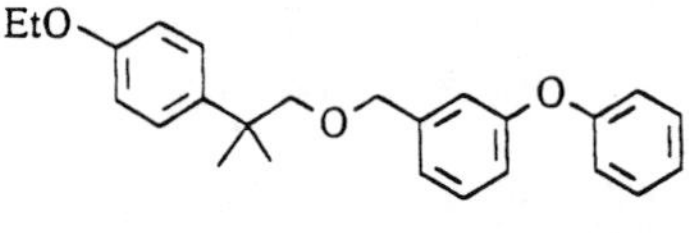

Etofenprox

Table 3   *Advantages of Pyrethroid Insecticides Claimed by Manufacturers*

1. Effective on a wide range of pests.

2. Active on all life-stages of insects, by cuticle and stomach uptake.

3. Rapid knockdown activity (reduces disease transmission).

4. Persistent protection (photostable).

5. Low use rates.

6. Low environmental impact.

7. More cost-effective than other insecticides.

8. Rain-fast formulations.

9. Compatible with most other products when mixed.

10. Proven yield benefits.

11. Less unpleasant smell than most other insecticides.

12. Suitable for ground and aerial application.

13. No phytotoxicity (plant damage).

Although some companies have succeeded in differentiating their proprietary product by 'premium' effects and yield benefits, pyrethroids have generally achieved commodity status, with 35 brands or more available in many countries. Generic producers have emerged particularly in Asia and eastern Europe, causing a steady erosion of prices. Recent expiry of the UK Government's NRDC patents will result in further generic manufacture. Pyrethroid products typically cost an American cotton grower about $12 to $15 per hectare, one half the price of most leading non-pyrethroid products. Season-long insect control usually represents 20% or less of his total variable costs of production. A cotton grower in India might spend 300 rupees per hectare for each pyrethroid application aimed at preventing bollworm caterpillars from damaging a crop worth 10,000 rupees per hectare. Benefit/cost ratios are thus high even when four or five sprays are made to combat prolonged pest infestations. It was recently estimated that the hypothetical substitution of pyrethroids with the best alternative chemistry, if say resistance became widespread or registrations were withdrawn, would increase global insect control costs by 39%, or approximately $2.4 billion[7].

The largest selling pyrethroids today are cypermethrin, fenvalerate, deltamethrin and lambda-cyhalothrin. Sales of each of these exceed $100 million per annum, although they have some way to go to match the versatile performance of the world's leading insecticide, chlorpyrifos, with sales in excess of $300 million. Table 4 highlights the diversity and geographic spread of crop and public health outlets for one pyrethroid, lambda-cyhalothrin, sold generally under the trade-name "Karate". The dominant markets for nearly all pyrethroids are foliage and fruit-feeding pests of cotton, vegetables and cereals, as well as public/animal health pests such as cockroaches, flies and mosquitoes. However, a few have established niches of their own, for example, tefluthrin has sufficient vapour pressure to be used as a soil-applied insecticide; bifenthrin is a potent acaricide, and etofenprox is a paddy rice insecticide. Recent introductions, such as zeta-cypermethrin and beta-cyfluthrin, have largely been based on greater cost-efficacy, usually through partial resolution of the isomers, rather than additional or novel pest spectrum.

Table 4   *Leading Markets for Karate Insecticide*

Crop	Pests	Countries
Cotton	Bollworms, armyworms bollweevil	USA, China, Pakistan, Brasil, Former Soviet Territories, Central America
Cereals	Aphids	France, Germany
Vegetables	Leaf-feeding caterpillars, aphids, thrips, beetles	Europe, S.E. Asia
Soybean	Leaf-feeding caterpillars, Pentatomid bugs	Brasil, Argentina, S.E. Asia
Fruit	Leaf- and fruit-feeding caterpillars, aphids	Spain, E.Europe, Turkey
Maize	Stem-boring and Ear-feeding caterpillars	South America, France
Vector Control	Mosquitoes	Middle East, Asia, South America
'Professional' Pest Control	Cockroaches, Termites	USA, Europe
Municipal Pest Control	Cockroaches, Flies	Middle East

4 PROSPECTS FOR CONTINUED SUCCESS

Partly as a result of the impressive sales record described above, pyrethroids face a number of biological and political challenges, the most notable of which is resistance (Table 5). Although resistance in pests **per se** rarely results in the withdrawal or total failure of any insecticide, it does erode efficacy over time, which encourages users to increase rates, mix cocktails of different chemistries, switch to newer products if available, or even to plant crops less dependent on chemical control. Whilst there are synthetic alternatives to overcome metabolic causes of resistance, including halogen substitution to prevent oxidative conversion to polar metabolites, or molecules not subject to ester cleavage, the most sensible approach is to delay the onset of resistance in the currently available ester compounds.

Table 5   *Documented Cases of Pyrethroid Resistance (Insecticide Resistance Action Committee of GIFAP, 1994)*

Outlets	*Pests**
Cotton	Aphids, whitefly, bollworms, armyworms
Fruit	Aphids, mites, psyllids, leaf miners
Vegetables and other field crops	Aphids, whitefly, leaf- and fruit-feeding caterpillars, Colorado Potato Beetle
Public Health	Houseflies
Animal Health	Various flies

*   It is not implied that resistance in these pests has occurred in all or even most countries. The spread and intensity of resistance varies with time, and it does not necessarily cause uneconomic levels of control.

It is easy to focus on the extreme examples of resistance induced by twenty or more sprays in one season, as in the cotton-growing areas of Thailand and southern India, or brassicas in South-East Asia where seemingly intractable problems have developed. However, it is also common for an initial crisis in control of one or more major pest species to promote adoption of co-operative management programmes, whereby all suitable insecticide groups are alternated or rotated, often on a regional scale to avoid undue selection pressure on any one mode of action.[8] Management programmes in Australia, USA and parts of Europe have been at least partially successful in delaying the onset of serious economic problems and in allowing time for alternative, integrated measures to be adopted. Similarly, in some public and animal health markets, restricting residual treatments of the more persistent products, in favour of non-persistent knockdown sprays, has enabled continued use of pyrethrins and pyrethroids when otherwise their usefulness would have been lost. The real challenge, especially in developing countries, is to educate growers and pesticide retailers in sensible use practices based on long-term goals, a joint responsibility of manufacturers and government agencies. Much more will be seen in future of concerted efforts to raise standards of insecticide usage amongst growers.

One potential side-effect of pesticide use is its impact on the environment, both aquatic and terrestrial. Laboratory toxicity data for fish and non-target arthropods can cast pyrethroids in a relatively unfavourable light, whereas in commercial practice, as extensive field studies have shown, any short-term hazard from recommended dosages is far less than for most other chemistries. Unwanted side-effects are invariably transitory (ie. weeks rather than months) in nature.[9,10] Nevertheless, the reputation of pyrethroids as incompatible with the goals of IPM and ICM has restricted their registration or recommendation in some major crops, notably rice and deciduous fruit.

A promising approach to minimise environmental and handler risk lies in novel formulations, such as micro-encapsulation, which can enhance selectivity between pests and predators and reduce exposure of users to concentrated active ingredients. Correct timing and placement of sprays can also serve to distinguish between friend and foe, which re-inforces the need for farmer education and improved application techniques to achieve safe and cost-effective control.[11]

Pyrethroid use has been associated with the phenomenon of "pest resurgence", whereby following an initial reduction in insect populations, they quickly multiply to levels greater than those seen in nearby untreated areas.[12] Clear evidence that the cause is destruction of natural enemies or direct stimulation of fecundity in pests by chemical deposits is lacking in most studies, but this undesirable side-effect of pyrethroids in some crops has entered the folk-lore.

In most other respects, the correct use of pyrethroids represents negligible risk to users, consumers and the environment. Low intrinsic toxicity to earthworms, birds and mammals, coupled with limited persistence, lack of mobility in soil and plants, and rapid metabolism by mammals and soil micro-organisms, is a profile possessed by few other insecticides and explains the long lists of registered uses in every continent.

A largely unpredictable challenge to certain traditional pyrethroid markets has emerged with the development of genetically engineered crops. If the protagonists of transgenic varieties, such as cotton expressing Bt-toxins, are to be believed, the antiquated system of atomising toxic chemical sprays over entire fields is soon to be replaced by subtle, built-in control measures, such as insecticidal proteins expressed in vulnerable parts of the plant only when switched on by environmental or physiological cues. A scenario in which major cotton and vegetable pests like *Heliothis* and *Spodoptera* are effectively controlled in this way, without resistance to toxins developing rapidly, would either necessitate the re-direction of pyrethroid use against those pests not affected by the currently-available toxins, or new registrations in crop and non-crop outlets not amenable to foreign gene technology. In reality, transgenic crops are likely to become established in the next decade as another weapon in the IPM armoury, for use alongside conventional chemicals, cultural practices and biological control agents, not as a panacea for all pest problems.

Pyrethroid sales will probably reach a plateau within the next five years, with an annual consumption in excess of 5,000 tonnes of cypermethrin equivalent, followed by a slow decline due mainly to competition from new types of natural and synthetic chemistry, and other pest control techniques. Healthy sales and profits from this outstanding group of insecticides are set to continue into the 21st century, supported by investments in improved delivery systems, especially more sophisticated formulations, complementary technologies, such as field kits for identifying pest species and their resistance mechanisms, responsible product stewardship and end-user education. A fitting testimony to the many chemists who diligently searched for synthetic alternatives to pyrethrum extract for more than half a century.

References

1. N. Morton and M. D. Collins, 'Pest Management in Cotton', ed. M. B. Green and D. J. de B. Lyon, Ellis Horwood, Chichester, 1989, Chapter 14, p. 155.
2. J. P. Leahey, 'The Pyrethroid Insecticides', Taylor and Francis, London, 1985.
3. Anon., *Reference Volume of the Agrochemical Service*, Wood Mackenzie, May 1993.
4. M. Elliott, N. F. Jones and C. Potter, *Annual Review of Entomology*, 1978, **23**, 443.
5. M. Elliott, A. W. Farnham, N. F. Jones, P. H. Needham, D. A. Pulham and J. H. Stevenson, *Proceedings of the 7th British Insecticide and Fungicide Conference (Brighton)*, 1973, **2**, 721.
6. G. G. Briggs, 'Advances in the Chemistry of Insect Control III', The Royal Society of Chemistry, Cambridge, 1994
7. I. A. Watkinson, *Pesticide Science*, 1989, **27**, 465.
8. P. K. Leonard and R. M. Perrin, *Proceedings of the Brighton Crop Protection Conference - Pests and Diseases*, 1994, **3**, 969.
9. C. Inglesfield, *Pesticide Science*, 1989, **27**, 387.
10. I. R. Hill, *Pesticide Science*, 1989, **27**, 429.
11. G. A. Matthews and E. C. Hislop, 'Application Technology for Crop Protection', CAB International, Wallingford, 1993.
12. M. R. Hardin, B. Penrey, M. Coll, W. O. Lamp, G. K. Roderick and P. Barbosa, *Crop Protection*, 1995, **14**, 3

Analysis of Exposure to Pesticides Applied in a Regulated Environment

A. J. Gilbert

PESTICIDES GROUP, CENTRAL SCIENCE LABORATORY (CSL), MINISTRY OF AGRICULTURE, FISHERIES AND FOOD (MAFF), HATCHING GREEN, HARPENDEN, HERTFORDSHIRE AL5 2BD, UK

1. INTRODUCTION

Pesticides are intentionally designed as agents enabled by their potential biological activity to be utilised to control certain species of living organisms in circumstances where those organisms are deemed to be pests. Their capacity for selectivity is a key criterion, ideally allowing specific pests to be controlled by the pesticide, while avoiding any harmful effect to all other species, be they crops, beneficial species or other non-target organisms. Of particular importance is the need to safeguard the health of those who work with pesticides who may become occupationally exposed to them.

Application of pesticides, a process usually intended to achieve their widespread distribution in finely divided form (typically as a liquid spray), is likely to present a risk of contamination to equipment operators. Thus the study of operator exposure to pesticides has long historical roots; a milestone being the publication in the Bulletin of the WHO of a protocol[1] that became an effective working standard for the direct measurement of operator exposure before a more formal WHO standard protocol was published[2]. By the adoption of a more standardised approach, results from early studies of applicator exposure to pesticides[3] accumulated a data base from which a better understanding of the routes of exposure and pattern of potential risk to workers due to pesticide exposure has been forthcoming.

1.1 Regulation of Pesticides

In order consistently to maintain the most favourable balance between potential costs and benefits of pesticides to society, their use has been subject to a system of regulation. Originally established upon a voluntary basis in the UK, this regulation moved onto a statutory basis under the Control of Pesticides Regulations (COPR) 1986. Current regulatory arrangements, which are becoming more complex with the inception of EU legislation,[4] are described in the MAFF/HSE 'blue book' Pesticides 1995,[5] which is issued annually. The regulatory framework is based on approval of products under stated conditions of use, that are to be used following approved label instructions (noting especially the 'statutory box'), according to Approved Codes of Practice, by operators with appropriate training and certification who operate equipment fit for its purpose.

Among the steps taken before approval can be granted is an operator risk evaluation. This may use a predictive modelling approach, such as the UK. Predictive Operator Exposure Model (POEM),[6] which is explained in greater detail in section 2. A mixture of product specific as well as generic data is required, the applicant for approval generally supplying the former and the regulatory framework itself providing the latter (i.e. standard reference points for outlining the envelope of likely practical conditions) that are necessary for input to the model. The background practical conditions may differ between different countries, such as Germany, UK and the USA or Canada, but many generic features of predictive modelling are common among the individual models in use.[7] Now a harmonised approach is being pioneered in the European Predictive Operator Exposure Model (EUROPOEM)[8,9] to provide an integrated framework for risk assessment to support European pesticide approval under the plant protection products directive. Geographical variations and alternative use conditions involved will test the flexibility and robustness of the harmonised model. New data, to complete the data base will be required, especially for protected crops and for re-entry exposure, and bridging studies may be necessary to test the validity of applying data collected from one region to support assessment of a product to be used in another place.

1.2 Management and Control of Hazardous Substances

To complement the safeguards derived from pesticide approval by regulatory departments, there is a general need for hazardous substances to be used under carefully managed circumstances, supported by an assessment of the local safety regime. The Control of Substances Hazardous to Health (COSHH) Regulations, issued under the UK Health and Safety at Work (HSW) Act[10] call for, inter alia, assessment of the exposure risk arising from hazardous substances used in the workplace, and for further monitoring and regular verification of compliance with published exposure ceilings, i.e. maximum exposure limit (MEL) or occupational exposure standard (OES) values, for certain cases. Section 2(2)c of the HSW Act requires employers to provide necessary information to employees, and UK Management of Health and Safety at Work (MHSW) Regulations[11] require employers to provide employees with comprehensible and relevant information about risks to their health and safety and about preventative and protective measures. Sound operational management to create and maintain safe working systems throughout the pesticide using industries is crucial to overall safety in practice.

1.3 Pesticide and Equipment Industries

Operator exposure data is required from the pesticide industry to facilitate risk assessment in support of product approval. In the UK the British Agrochemicals Association (BAA) has commissioned such a study[12] to contribute generic data to the POEM.

There is also a need for testing in its own right of the equipment that is used to apply pesticides. A survey of safety features of crop sprayers[13] identified the importance of protective cabs, devices for handling of concentrated pesticides and automatic booms among the systems necessary, in addition to protective clothing, for control of operator exposure to pesticides. Current regulatory impetus for safe systems of applying pesticides is now coming from within the legal framework of the European Union (EU), for example prEN 907[14] which gives standard safety requirements for sprayers and liquid fertiliser distributors used in agriculture and forestry machinery. Those requirements may include the fitting of a transfer system for pesticides to protect operators from risks of exposure. A standard specification for design, performance and testing of induction hoppers[15] and for closed transfer of liquid formulations,[16] in turn gives details for alternative means to meet the required standard of performance. Instructions for safe operation of equipment, which must be fit for the purpose it is used for, must be available to users under the provisions of the HSW Act. Studies for the purpose of the testing of correct function of parts, the sensible design of systems and the evaluation of foreseeable adverse effects in order to provide precautionary advice to users are all therefore necessary.

To meet an increasing overall demand in the future for pesticide application safety data, particularly that for worker exposure, there is need for an integrated and internally consistent regulatory framework for use by authorities, industry, users and advisors alike, throughout the range of international and national organisations concerned with worker safety in use of pesticides. The increasing sophistication of pesticide application technology, and of the methodology for study of associated occupational exposure for worker risk evaluation, at national and local level, is likely to increase both the number and complexity of studies needed to be carried out. The quality and usefulness of such data will depend on proficiency in the conduct of analytical science concerned, but also on awareness of the underlying framework of factors that describe and control the manner of pesticide use, and the work practices being scrutinised.

2. PREDICTIVE APPROACHES TO RISK ASSESSMENT

Operator risk evaluation in the UK has been based upon the Predictive Operator Exposure Model (POEM), while other EU countries, such as Germany, have their own model[17] and currently a concerted action for harmonisation of predictive worker exposure modelling within the EU is underway with the aim of developing a European Predictive Operator Exposure Model or 'EUROPOEM', which incorporates the best features and data from individual models used in different countries.

The UK POEM illustrates fundamental background common to predictive exposure models; for example fig 1 shows possible routes of exposure. Measurements for operator exposure and potential exposure arising by these different routes are obtained using appropriate study methodology, which is explored in more detail in the next section. The diversity of alternative ways of using individual pesticides, together with the generic nature of application practices and the overall complexity of factors that may control pesticide contamination (and hence exposure) make it advantageous to use a predictive modelling approach to quantify potential exposure, hence to evaluate risk. The exposure

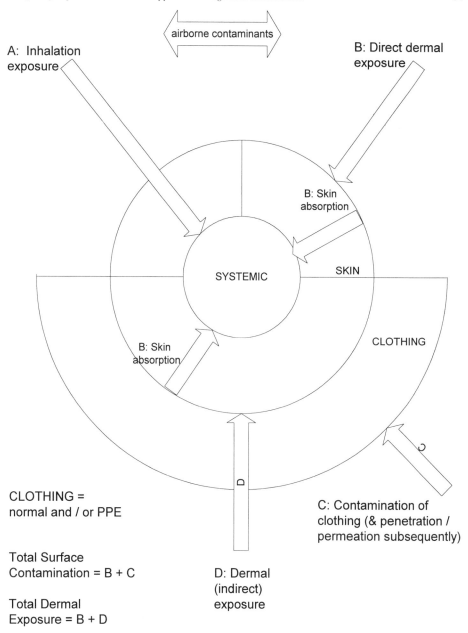

Figure 1. Alternative routes for operator exposure to pesticides.

(After Martin, 1990)

data must account for the mass of active ingredient (a.i.) to which the operator is exposed per unit mass handled or applied. Ideally, exposure should also be possible to express as a proportion of the applied volume of diluted preparation, because volume rate is a factor in selecting application method, which may alter the level of exposure. The distribution pattern of exposure to different parts of the body, which again may depend on application method, needs to be found. Separate values for inhalation exposure and dermal exposure should be available and the latter separable into the major regions of head, hands, body and legs. The exposure (or at least relative total surface contamination) levels arising from different application methods should be possible to differentiate. The exposure arising from handling the concentrated product should be possible to distinguish from that arising from applying the prepared mixture.

The predictive modelling approach facilitates combination of data specific to individual cases (e.g. physico-chemical properties of the a.i.) with that generic data which is representative of the case (e.g. operator total surface contamination from the application or handling process). The precise nature of each type of data and the exact demands of each case determine which data may contribute to an exposure evaluation that validly affords the necessary level of assurance in the outcome of the overall risk assessment. Hence EUROPOEM is typical of models in being able to support evaluations at varying levels of specificity in their data mix, all leading to a three tiered system of risk assessment as explained by Fig.2. Thus many cases could be easily and rapidly assessed at the first tier, making the most use of generic data and using conservative assumptions (with a high inherent safety margin) for unknown factors, but the assessment can operate at the second or third tier for cases that warrant a more complex assessment and call for more product specific data, including information that may be possible to obtain only from field experiments dedicated to the case. The key values that are needed for the risk assessment are the predicted exposure level (PEL), which is derived from the model and the no-observable-adverse-effect-level (NOAEL), which is derived from the intrinsic toxicology of the pesticide. An acceptable operator exposure level (AOEL) can be set which is less than the NOAEL and consistently greater than any likely PEL values for the pattern(s) of use of the product, as long as an adequate margin of safety (MoS) exists due to the worker exposure in the circumstances of using the product being at most a small fraction of a dose known to be below the level where any adverse effect could be expected.

Uniform principles for risk assessment of pesticides under the Plant Protection Products Directive (91/414/EEC) call for risk assessment for workers and bystanders. A distinction also needs to be made between those who work directly with the pesticide ('operators') and others who work in jobs carrying an indirect risk of exposure to pesticides, such as harvesters who may be subject to 're-entry' exposure. All of the above classes of workers are distinct from 'bystanders', which the EUROPOEM defines as persons located within or directly adjacent to the area where pesticide application or treatment is taking place; whose presence is quite incidental and unrelated to work involving pesticides but whose position may put them at risk of potential exposure (a worst case as there is little scope for assured control measures to prevent exposure) and for whom it is assumed that no protective clothing is to be worn and perhaps little ordinary clothing. CSL has for over 20 years carried out routine field measurement of

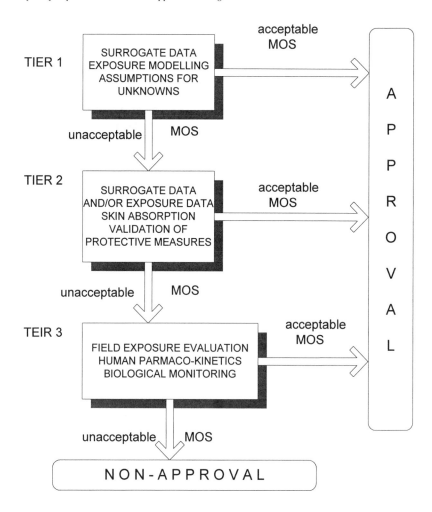

Figure 2. General scheme of a three tier approach for the assessment of worker exposure in a risk assessment procedure for registration.

Note: requirement of actual human data (field exposure evaluation, biological monitoring) depends on the estimated margin of safety (MOS), i.e. the ratio between (surrogate) exposure data and toxicological no-observed-adverse-effect-level (NOAEL).   (After Henderson et al 1993) spray drift which includes volunteer human subjects as 'Bystanders' within the standard array of drift targets;[19] thus it has accumulated experience and an extensive background database in this area.

For operator exposure field assessments a distinction should be made between operators who only apply the mixture ('applicators'), those who only mix and load the pesticide ('mixer loaders') and those who do both jobs ('mixer loader / applicators'). For applicator studies a further system is necessary to discriminate between different sorts of equipment, for example sprayers producing different qualities of spray,[20] and techniques e.g. low vs high level spraying (i.e. above or below the operators waist level). The simplest criteria for separate categories common to different models[21] comprise differentiation for field crops vs. bush crops vs. tree crops (orchards), for tractor mounted/trailed vs. hand held equipment and for indoor vs. outdoor operations. Categorisation needs to be pragmatic and serves best by offering an objective basis for describing systematically any combination of equipment and technique, so enabling respective exposure data sets to be compared only on a like vs. like basis and so validly differentiate between cases. A scheme for classification of equipment and techniques by hazard in its own right,[22] i.e. from the physical hazard of loss of control over applied material, as opposed to the biological hazard intrinsic to the pesticide, is an essential tool. It must enable all possible alternative practices (including innovations) to considered in all their subtle complexity and yet retain the ability to grade them on a simple scale (e.g. no more than four classes) that is understandable by users selecting appropriate means to apply pesticides of any given hazard category. For mixer loader and mixer loader / applicators studies, involving filling and transfer, the context provided by developing standards for design and performance of closed transfer systems for pesticides and induction bowls should be taken into account. Manual pouring into sprayer tanks can be studied directly, but variable factors such as pouring technique, container size and formulation type may make comparison of data difficult. Laboratory study of operator contamination and spillage from manual pouring of liquid pesticides has been standardised according to the protocol routinely used by CSL for such studies,[18] which underlies the extensive UK database in this area.

To illustrate not only the need for, but also the difficulties found with setting appropriate distinctions between practical scenarios, such as described above, the example is given of study into operator exposure arising from pesticide spraying from all-terrain vehicles (ATV). This innovative and atypical practice was expected to increase operator exposure risk. That in turn highlighted the role that should be played by industry to examine, qualify and quantify the risk to operators contingent upon use of the equipment or components it supplied, as well as that of enforcement authorities to oversee this process in a meaningful and objective way. The next question to arise, however, was what exactly was meant by an 'ATV', or even an 'atypical' vehicle. A consultation with industry produced a definition of an ATV[23] which included not only the four wheel drive motorcycles that were intended to be covered, but also encompassed the much wider range of vehicles all suited to travelling 'off road'. Notwithstanding queries over the precise definition of ATV, a limited field study of a commercial application of pesticide by ATV was carried out by HSE[24] which suggested that operator exposure when using the ATV was likely to be higher than when using conventional tractor mounted equipment. However in that study the spray applicator (ATV driver) also did the mixing and loading so the measured operator exposure data included the concentrate handling, which could then not be ruled out as a source of contamination. So the safety case for or against spraying by ATV per se could not be objectively tested because the exposure arising from handling the concentrate, which would probably exceed that arising during field spraying by boom and nozzle from most kinds of vehicle including the ATV, could not be differentiated from that arising only from the application, and dependent on the 'atypical' nature of the ATV. Subsequently a further study[25] taking atypical vehicles (e.g. ATV) as its subject matter demonstrated the utility of whole body dosimetry and artificial tracers (visible dyestuffs) as a safe and convenient means for assessing comparative contamination risks arising to operators from alternative types of application equipment. Contamination not exposure was measured and the protocol itself was the subject of the test, hence the working variables chosen for the ATV were not representative of such equipment as typically used, but were deliberately set to be extremes of feasible practice. Data from that study showed critical application variables such as spray quality, boom height and direction of travel with respect to the prevailing wind direction matter greatly in determining comparative exposure risk, both in combination and in isolation (although the validity of considering them in isolation may be questioned). While product specific data from 'third tier' experimental field studies has great value in itself, the ATV examples show how predictive exposure modelling can serve as an essential aid in interpretation of the findings of any single study.

3. ALTERNATIVE METHODS FOR ASSESSMENT OF EXPOSURE

Alternative methods for assessment of worker exposure to agricultural pesticides have recently been subject to rigorous review at an international workshop held in the Hague.[26] It was recommended that modern field studies should be carried out in accordance with respective agreed guidelines, whether generally evaluating agricultural worker exposure to, and absorption of, pesticides,[27] or attempting biological monitoring of a particular pesticide.[28] The guidelines have the advantage of being sufficiently flexible to accommodate various objectives and monitoring methods, so are equally applicable to all three separate categories of operator exposure data (i.e. applicator, mixer loader and mixer loader / applicator), as well as for re-entry workers and bystanders, and do not substitute for the detailed planning needed within individual study protocols or their component standard operating procedures. However, the guidelines ensure that studies can conform to a common sound basis, that all data are validly gathered and verified and that conclusions drawn from data are truly supported by the design of the study. Important ethical issues are discussed as well as the need for studies to identify and qualify separately realistic worst cases, typical industrial situations and experimental situations, while avoiding 'careful worker syndrome' and other general causes of bias in results.

All studies share a common need to select carefully the units for expressing exposure results and setting these against an appropriate base. Generally the exposure value is expressed as a ratio where the numerator corresponds to either the absolute mass

Table 1. Options for units to express exposure data (in terms of test substance)

Dermal exposure	mg (bodypart)$^{-1}$, mg day^{-1} whole body, mg (kg body weight)$^{-1}$ day^{-1}, mg kg^{-1} test material handled or applied
Inhalation exposure	μg m^{-3}, μg day^{-1}, μg (kg bodyweight)$^{-1}$ day^{-1}, μg kg^{-1} test material handled or applied
Absorption	mg day^{-1}, mg (kg bodyweight)$^{-1}$ day^{-1}

of exposure or that mass as a proportion of the mass or volume of pesticide handled or applied, while the denominator signifies the overall rate of the exposure process either simply as time, or by reference to some pertinent work related variable (e.g. job completed, area treated, throughput of treated mass, etc.') or environmental variable such as air volume, with respect to breathing rate, work activity and air exchange regime at the sampling point. It is highly advantageous to be able to translate results to alternative bases to enable inter comparison of results between studies and for overall collation of data into modelling frameworks. Not all types of exposure measurement, which are described below, offer the same scope for adapting the units in which data are expressed.

3.1 Ambient Monitoring

Ambient monitoring enables quantification of total surface contamination and/or total dermal exposure, depending on the sampling strategy, and of inhalation exposure. Alternative sampling strategies may be selected, but ultimately all lead to an estimation of exposure levels within the body because the percutaneous penetration and absorption of dermal contaminants is critical in determining the systemic exposure level. Three key terms are used to determine how combinations of sample and measured analyte correspond to direct or indirect exposure by different routes. Potential dermal exposure is the total amount of pesticide coming into contact with the protective clothing, work clothing and the skin. Actual dermal exposure is the amount of pesticide coming into contact with bare (uncovered) skin plus the fraction penetrating protective and work clothing to the underlying skin, and which is therefore available for percutaneous absorption. The 'inspirable fraction' of airborne particulate is that fraction capable of entering the respiratory tract via the nose and the mouth, so providing a source of absorption into the body, either from direct inhalation or from secondary oral ingestion.

Alternative approaches to dosimetry govern how directly the dermal exposure can be estimated. Contamination levels can be measured on the outside of all clothing, between protective clothing and work clothing (worn underneath) or on the skin surface itself. The protective efficiency of the protective clothing and of work clothing in use will affect the relationship between these alternative measurements.[29, 30] Patches of known size can be affixed to inner or outer surfaces of clothing or to skin, or disposable clothing may be used for whole body dosimetry.[31] Values from patches need to be extrapolated upwards to estimate exposure to the whole body, while whole suits would need sectioning to allow the distribution of total contamination to be evaluated. The pattern of distribution of contamination is likely to be non-uniform and vary to different degrees for different occupational exposure scenarios,[32] so the sizing and placing of patches, or sections into which whole suits are cut may be critical factors that determine the eventual accuracy of exposure estimates. All sample matrices should be known to have an adequate capacity to retain all of the analyte that is to be caught, and must be subject to careful control procedures to verify their recovery factor. All matrices should be subject to fortification with known volumes of the pesticide or tracer mixture applied in the field to validate the results, and allow for systematic losses through potential degradation which could occur between the start of a trial and the end of laboratory analysis. Selection of suitable tracers should take such factors into account.

Pesticides may be substituted by synthetic tracers such as visible or fluorescent dyestuffs, as long as the simulant is known to behave in a manner truly representative of the pesticide. This can have safety advantages if the potential exposure consequences of alterations in pesticide use practice e.g. choice of personal protective clothing (PPE), are to be investigated without putting workers into situations of actual risk. The PPE example illustrates where tracers and real pesticides may or may not be interchangeable, insofar as both may act similarly in solution to penetrate the clothing (i.e. through holes, fastenings or gaps) but may not mimic each others permeation characteristics. European standard (EN) test methods for resistance of chemical protective clothing materials to permeation and penetration by liquids[33, 34] may be employed to determine the effective similarity between pesticides and alternative substances used as tracers in liquid solutions. The direct measurement by surface fluorimetry of dermal deposits of both liquids[35] and of dusts[36] is rapidly evolving as a convenient, accurate and versatile technique. It has the advantage of allowing direct measurements from the skin surface in a non-invasive manner and can be employed in addition to biological monitoring of absorption of actual pesticide to obtain corroboration between ambient and biological monitoring as well as a much fuller picture of the internal dynamics of contamination and exposure processes. A laboratory study[37] of high level spraying in a confined space compared results from simultaneous assessment using traditional chemical extraction of both pesticide and fluorescent tracer from sectioned disposable suits, surface fluorimetry of deposits on the skin and biological monitoring of pesticide metabolite in urine. Results from all techniques were seen to be in close agreement, both among themselves and with the original predicted estimates of contamination and exposure derived from modelling, although the trial was too small to allow rigorous statistical analysis.

3.2 Biological Monitoring

Biological monitoring (BM) is the measurement of a pesticide or its metabolites in the body fluids of exposed persons and conversion to an equivalent absorbed dose of the pesticide based on a knowledge of its human metabolism and pharmacokinetics. Measurements may be of a.i or metabolite within the tissues, secreta, excreta or expired air. Many sorts of exposure studies may be described as BM, especially (but not exclusively) if the sampling matrix is 'biological' (breath, urine or blood), but assay of a toxic substance in air, for example, can be included.

It is tempting to imagine that BM is a straightforward way to assess exposure, measuring the substance in question in samples taken directly from the human subject, but a great many factors can conspire to make this sort of exposure evaluation technically challenging and difficult to validate. However, such an assessment when successful has great value and can be called for within the regulatory framework, either to investigate or to monitor exposure to hazardous substances. In order to safeguard workers welfare such study may be carried out in close association with other health monitoring, so there is scope for confusion of BM with other health related study objectives (especially with BEM - see below) and the meaning of results may require careful interpretation. Due to the need for clarity in matters of safety regulation, in the UK the Advisory Committee on Toxic Substances has developed a strategy for biological monitoring,[38] which sets out clear definitions. Similarly the HSE publish official guidance for biological monitoring in the workplace,[39] which is currently undergoing revision to include a template of key criteria for establishing a biological monitoring procedure.[40]

3.3 Biological Effect Monitoring

Biological effect monitoring (BEM) is the measurement and assessment of early non-adverse biological changes caused by absorption of the substance in exposed workers. Exposure may be estimated from measured alteration of an indicator system (e.g. Acetyl Cholinesterase activity), but without necessarily assessing 'health' consequences. Otherwise study of exposure together with related health effects by epidemiological investigation may be undertaken.

It is notable that BM and BEM are often included under the umbrella heading of "biological monitoring". Three further concepts from the developing strategy for biological monitoring are important to consider so that alternative studies, having different objectives in respect of exposure evaluation, can contribute data appropriately toward a common system for safeguarding workers health. This applies especially to BEM studies that attempt more directly to relate exposure with health assessment. First there may be a biological action threshold (BAT), which represents an exposure level which would prompt remedial action of some kind to lower exposure. Second there may be a health guidance value (HGV), which relates biological exposure levels of substances to potential health effects, although does not necessarily imply the need for immediate action. Thirdly there may be a benchmark value (BV) which indicates biological levels of a substance based on what is reasonably expected to be found within a normal industrial setting. It is acknowledged that setting of BAT levels is notoriously difficult given the inherent variabilities both in the working environments, practices and processes under study (i.e. that would also affect ambient monitoring) and in the biological variability within the human population in absorption, metabolism, distribution and excretion of a substance and in pathological response to any given systemic dose received.

4. DISCUSSION

This single paper cannot, and does not attempt to cover the multitude of detailed considerations that are necessary to include in the planning, conduct and reporting of any single study. It intends to provide a broad perspective of the background regulatory framework, together with pertinent practical and scientific issues, and to raise questions that address the implications of decisions that must be made when conducting analysis of worker exposure to pesticides. Sources of detailed guidance are given and in many cases they will, in turn, refer to other standards or conventions. It is recommended that enquiries made prior to exposure assessment are as exhaustive as possible in order to identify component factors that should be set, controlled or measured to support the interpretation of study findings. The measurements that are to be made and methods for making them should be chosen to suit the exposure evaluation objectives. The requirement for exposure evaluation data are set by the needs of the risk assessment, and this should guide the approach taken to relevant studies. Predictive modelling relevant to the case should be explored thoroughly before field study to define the purpose(s) for which the data can be used (and can be made most useful). Ultimately, the comprehensive safety evaluation and management system should be addressed. Adherence to safety practice during experimental studies must always be maintained (e.g. compliance with COSHH, following approved label instructions and correct use of equipment) and hence the scope of new data to support specific changes to products or practices should be firmly established.

References

1. W F Durham & H R. Wolfe, Measurement of the exposure of workers to pesticides. *Bull Wld Hlth Org*, 1962, **26,** p 75..

2. WHO. Field surveys of exposure to pesticides. Standard protocol. Document VBC/82.1, 1982 World Health Organisation (WHO), Pesticide Development and Safe Use Unit, Division of Vector Biology and Control, WHO Headquarters, Geneva.

3. H R Wolfe, W F Durham & J F Armstrong. Exposure of workers to pesticides. *Arch Environ Health,* 1967, Vol **14,** p 622.

4. Commission of the European Communities (CEC). Proposal for a Council Directive establishing annex VI Directive 91/414/EEC concerning the placing of plant protection products on the market. OJ No L 230 19.081991, 1993 p1.

5. Pesticides 1995 Pesticides approved under the Control of Pesticides regulations 1986. Reference Book 500 Ministry of Agriculture, Fisheries and Food (MAFF) / Health and Safety Executive (HSE) 1995, HMSO, Publications Centre, P O Box 276, London SW8 5DT.

6. A D Martin A Predictive Model For The Assessment Of Dermal Exposure To Pesticides. In Prediction of Percutaneous Penetration. Methods, Measurements, Modelling. 1990 (Edited by Scott R C, Guy R H and Hadgraft J). IBC Technical services Ltd, Southampton.

7. J J van Hemmen Predictive exposure modelling for pesticide registration purposes. *Ann. Occ. Hyg,*. 1993, **37,** p 541.

8. EUROPOEM The development, maintenance and dissemination of a European Predictive Operator Exposure Model (EUROPOEM) Database. A Concerted Action under the AIR programme 1991 - 1994. Contract Number CT9... (PL No 921370).

9. W Chen & P Watts, European Operator Exposure Database. Health Canada and NATO Workshop on Methods of Pesticide Exposure Assessment, Ottawa 5-8 October 1993.

10. The Health and Safety at Work Act (1974). HMSO, Publications Centre, P O Box 276, London SW8 5DT

11. The Management of Health and Safety at Work Regulations (1992). HMSO, Publications Centre, P O Box 276, London SW8 5DT

12. Spray Operator Safety Study 1983. British Agrochemicals Association (BAA) Ltd. Alembic House, 93 Albert Embankment, London SE1 7TU.

13. D J Left, An evaluation of safety features on agricultural crop sprayers. Unpublished Report 1986, H M Agricultural Inspectorate, Health and Safety Executive Stanley Precinct, Bootle, Merseyside L20 3QZ

14. Draft British Standard. Safety requirements for agricultural and forestry machinery - sprayers and liquid fertiliser distributors (prEN 907). Document 92/84207 DC 1993, British Standards Institute (BSI) Committee on Agricultural and Garden Equipment (AGE/15) 389 Chiswick High Road, London W4 4AL

15. Draft British Standard. Spraying equipment for crop protection. Part 8 Specification for Induction Hoppers. Document 95/702596 DC 1995, British Standards Institute (BSI) Committee on Agricultural and Garden Equipment (AGE/15) 389 Chiswick High Road, London W4 4AL

16. Draft British Standard. Spraying equipment for crop protection. Part 9 Specification for closed transfer of liquid formulations. Document 95/702513 DC 1995, British Standards Institute (BSI) Committee on Agricultural and Garden Equipment (AGE/15), 389 Chiswick High Road, London W4 4AL

17. J-R Lundehn, D Westphal, H Kieczka, B Krebs, S Löcher-Bolz, W Maasfeld and E-D Pick, Uniform principles for safeguarding the health of applicators of plant protection products. Mitteilungen aus der Biologischen Bundesantalt für Land- und Forstwirtschaft, Heft 277, Berlin, Germany 1992.

18 G A Lloyd, G J Bell and A R C Dean, Operator Protection Group laboratory test for the measurement of spillage on pouring from containers of liquid pesticides. AHU Report Number 686, Pesticides Group, CSL (unpublished) 1982.

19. A J Gilbert and G J Bell, Evaluation of the drift hazards arising from pesticide spray application. 1988 Aspects of Applied Biology, **17** (Environmental aspects of applied biology).

20. S J Doble, G A Matthews, I Rutherford and E S E Southcombe. A system for classifying hydraulic nozzles and other atomisers into categories of spray quality. British Crop Protection Conference - Weeds 1985. BCPC.

21. P Y Hamey. A Harmonised European Model for Predicting Operator Exposure to Pesticides. Prediction of Percutaneous Penetration 1993, Vol. **3b**, Ed. K R Brain, V J James & K A Walters, STS Publishing, Cardiff, UK.

22. C S Parkin, A J Gilbert, E S E Southcombe and C J Marshall. The British Crop Protection Council scheme for the classification of pesticide application equipment by hazard. *Crop Protection*, 1994, **13**, **(4)** p 281.

23. Classification of A.T.V.s. Pesticides Application Technology Panel Paper PATP 12/5, July 1990. MAFF, Nobel House, 17 Smith Square, London SW1

24. E M Widdowson. Exposure of spray operators to pesticide and carbon monoxide during pesticide application from an all-terrain vehicle. Unpublished Report 1989, H M Agricultural Inspectorate, HSE, Great Eastern House, Tenison Road, Cambridge BC1 2BS.

25. A J Gilbert. Development of protocol for assessing potential operator contamination by pesticides when applied by specialist vehicle mounted / trailed equipment. CSL Contract Report C900484 1991, for Health and Safety Executive (H M Agricultural Inspectorate), Daniel House, Stanley Road, Bootle, Merseyside.

26. P Th Henderson, D H Brouwer, J J G Opdam, H Stevenson and J Th J Stouten. Risk Assessment for Worker Exposure to Agricultural Pesticides: Review of a Workshop. *Ann. Occ. Hyg,* 1993, Vol **37**, **(5)** p 499.

27. G Chester. Evaluation of agricultural worker exposure to, and absorption of, pesticides. *Ann. Occ. Hyg*, 1993, **37**, **(5)** p 509.

28. B H Woollen Biological monitoring for pesticide absorption. *Ann. Occ. Hyg,* 1993, **37**, **(5)** p 525.

29. R A Fenske Comparative assessment of protective clothing performance by measurement of dermal exposure during pesticide applications. *Appl Ind Hyg,* 1988, Vol **3**, **(7)** p 207.

30. T Thongsinthusak, R K Brodburg, J H Ross, D Gibbons and R I Krieger. Reduction of pesticide exposure by using protective clothing and enclosed cabs. Paper No 126, 199th Am. Chem. Soc. National Meeting, Agrochemicals Division, Boston, MA, April 22-27, 1990.

31. G A Lloyd, G J Bell, J A Howarth and S Samuels O. P. R. G. procedure for measurement of the potential exposure of spray operators to pesticide sprays - direct

method. Operator Protection Research Group (OPRG) information sheet No 20, 1986 CSL, Hatching Green, Harpenden, Herts AL5 2BD.

32. R A Fenske. Nonuniform dermal deposition patterns during occupational exposure to pesticides. *Arch. Environ. Contam Toxicol,* 1990, **19**, p 332.

33. BS EN 369 Protective clothing - Protection against liquid chemicals - Test method: Resistance of materials to permeation by liquids (1993). British Standards Institute (BSI) 389 Chiswick High Road, London W4 4AL

34. BS EN 368 Protective clothing - Protection against liquid chemicals - Test method: Resistance of materials to penetration by liquids (1993). British Standards Institute (BSI) 389 Chiswick High Road, London W4 4AL

35. R A Fenske, Correlation of fluorescent tracer measurements of dermal exposure and urinary metabolite excretion during occupational exposure to malathion. *Am. Ind. Hyg. Assoc J,* 1988, **(49)** p 438.

36. M W Roff A fluorescence monitoring system to measure dust concentrations on surfaces: Discussion of instrumentation and techniques. Report No IR/L/DF/89/5 (1989), Project No R48.49.MD, Health and Safety Executive Research and Laboratory Services Division, Broad Lane, Sheffield S3 7HQ.

37. M W Roff and J Cocker Comparison of fluorescence monitoring with biological monitoring and chemical extraction in simulated field spraying trials. Report No IR/L/DF/93/9 (1993), Project No R48.49.HPD, Health and Safety Executive Research and Laboratory Services Division, Broad Lane, Sheffield S3 7HQ.

38. A Strategy for Biological Monitoring (ACTS 43/94) HSC Advisory Committee on Toxic Substances (ACTS), 54th meeting, 9th November 1994.

39. Biological monitoring for chemical exposure in the workplace. Guidance Note EH 56 (Environmental Hygiene series) (1992), Health and Safety Executive, HMSO, Publications Centre, P O Box 276, London SW8 5DT

40. Template for establishing a biological monitoring procedure. Committee on Analytical Requirements (CAR), Working Group on Biological Monitoring (WG3) Paper Ref No CAR WG3/001/93, 1993

The Assessment of Operator Risk from Sheep Dipping Operations Using Organophosphate Based Dips

B. P. Nutley,[1] H. F. Berry,[1] M. Roff,[1] R. H. Brown,[1] K. J. M. Niven,[2] and A. Robertson[2]

[1]HEALTH AND SAFETY EXECUTIVE, HEALTH AND SAFETY LABORATORY, BROAD LANE, SHEFFIELD S3 7HQ, UK

[2]INSTITUTE OF OCCUPATIONAL MEDICINE, 8 ROXBURGH PLACE, EDINBURGH EH8 9SU, UK

1 INTRODUCTION

Sheep rearing is an important component of agriculture in the United Kingdom (UK). There are some 17 million sheep producing over 20 million lambs. The total value of meat and wool to UK farmers is estimated to be over £750 million a year[1]. However, sheep are prone to attack by a number of insect pests including sheep scab mite, blowflies, ticks and keds which cause painful and debilitating conditions. The conventional method of control is to dip sheep in a dilute insecticide solution in spring prior to these pests becoming a problem. The insecticide persists in the sheep's fleece for the majority of the summer protecting the animal from attack. If necessary sheep are also dipped in the autumn; this second dip extends protection until the onset of cold weather halts insect activity.

Insecticides used for dipping sheep must have suitable persistence. In the 1970s the organochlorine pesticide lindane (γ-hexachlorocyclohexane) was used. This was withdrawn in 1984 because of concerns about long term effects upon the environment from this class of insecticides. Lindane was replaced by several organophosphate ester pesticides of which only two, diazinon and propetamphos, are now licensed for use; chlorfenvinphos was also used until the end of 1994. Organophosphate esters were seen as suitable succesors to organochlorines because they have high insecticidal activity but only limited environmental persistence.

Organophosphate esters are readily metabolised by a variety of enzymes systems to polar compounds and excreted. An important metabolic pathway for organophosphate pesticides involves hydrolytic cleavage which leads to the formation of an alcohol and an acid moiety[2]. The alcohol moiety may undergo conjugation with glucuronic or sulphuric acid to form highly polar conjugates which are readily excreted.The acid moiety usually contains the phosphorus atom present in this class of pesticides and for many organophosphate pesticides this metabolite may be a dialkyl phosphate where, depending on the structure of the parent pesticide, the alkyl groups are either methyl or ethyl chains. The most important of these potential metabolites are O,O-dimethyl phosphate (DMP), O,O-dimethyl phosphorothioate (DMPT), O,O-dimethyl phosphorodithioate (DMPDT) and the ethyl homologues O,O-diethyl phosphate (DEP), O,O-diethyl phosphorothioate (DEPT) and O,O-diethyl phosphorodithioate (DEPDT). The structure of these metabolites is shown in Figure 1 along with O,O-dibutyl phosphate (DBP). The acid metabolites are often excreted without undergoing further metabolism.

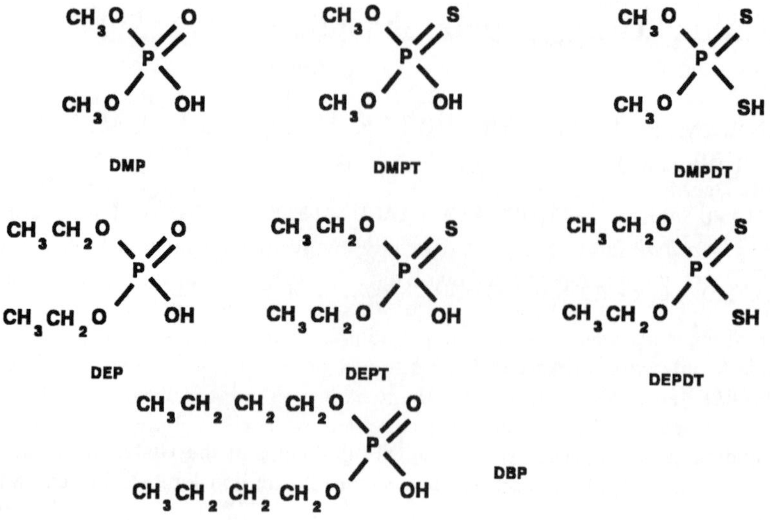

Figure 1 *Structure of potential dialkyl phosphate metabolites and internal standard*

The mechanism of action of the acute toxic effects of organophosphate ester pesticides involves inhibition of serine esterase enzymes[3]. Toxicologically the most important of these are believed to be acetylcholinesterases located in the nervous system at cholinergic nerve endings, e.g. neuromuscular junctions. Inhibition of acetylcholinesterase prevents hydrolysis of the neurotransmitter acetylcholine to choline and acetic acid. This leads to a build up of the neurotransmitter which prevents repolarisation of the nerve ending and inhibits transmission of further impulses across the junction. Inhibition of nerve endings brings about a variety of physiological changes that can, if the dose received is high enough, lead to death.

Acute effects of sub-lethal doses of organophosphate pesticides in man are well known and include sweating, salivation, abdominal cramps, vomiting, incontinence, muscular weakness, and breathing difficulties[4]. In recent years concern has also been expressed about long term effects following acute exposure to organophosphate pesticides. Research suggests that following an acute exposure some victims may show reductions in performance of certain neurobehavioural tests when tested several months later[5]. However, there are also concerns that people who do not appear to have suffered acute poisoning have developed debilitating illnesses following use of organophosphate pesticides. The symptoms described include extreme exhaustion, mood changes, memory loss, depression and severe muscle weakness. In the UK, these symptoms have been most frequently reported with farmers using organophosphate based sheep dips.

Sheep farmers may come into contact with organophosphate pesticides used for sheep dipping in a number of different ways. These include diluting the commercially available concentrate to prepare the dip bath, dipping sheep and handling the animals subsequent to dipping. Routes of exposure may potentially include inhalation of the dip (either as a vapour or as aerosols when the wet sheep shake themselves), dermal absorption following skin contamination and ingestion (e.g. smoking or eating without washing hands). Between 1991 and 1993 the UK Health and Safety Executive instigated research to investigate sources of exposure during the dipping process. This paper describes some these studies and the results obtained.

2 MATERIALS AND METHODS

2.1 Study Details

2.1.1. Principal Study. The principal study involved an occupational hygiene survey of 14 different dipping operations involving 38 workers using a variety of different designs of dip bath and looking at work practices of the personnel involved in dipping. The aim of the study was to investigate how dipping was actually performed; no attempt was made to influence workers to use additional protective measures to those that they would normally use. Two occupational hygienists were present throughout the dipping procedure at each site. Participants were classified as chuckers (throwing sheep into dip bath), paddlers (ensuring sheep were completely submersed) or handlers (rounding sheep up before and after dipping). The occupational hygienists noted the activities of all those involved in the dipping operation and scored each person for the extent of exposure during preparation of the bath, dipping and handling of sheep post-dipping. Farmers involved in this study used commercially available products licensed for use in the UK. The active organophosphate pesticide ingredients were diazinon and chlorfenvinphos although propetamphos was used on one farm. A full report of the findings of this occupational survey has been produced[6].

An additional procedure was used, on six farms, to assess exposure to dilute sheep dip by an alternative method to observation by the hygienists. This involved the addition of a fluorescent marker, at a known concentration, to the diluted dip and assessment of the extent of fluorescence on the skin of participants when viewed under ultraviolet light. Further details are given below.

In addition to the hygiene and contamination assessments biological monitoring was used to assess the extent of absorption of the organophosphate pesticides present in sheep dip. This involved obtaining urine samples from workers prior to dipping (pre-dip), at the end of dipping (post-dip) and prior to starting work the day after dipping (next day). Thirty seven of the 38 workers agreed to provide urine samples. These samples were analysed for urinary dialkyl phosphates (for details see below). Of the organophosphate pesticides used in sheep dip products diazinon forms DEP and DEPT whilst chlorfenvinphos only forms DEP. The assay is not suitable for monitoring exposure to propetamphos but urine samples from the one site using a product containing propetamphos were included for comparison.

2.1.2 Second Study. The second study involved an assessment of exposure to organophosphate pesticides via inhalation during sheep dipping operations. This assessment involved a contractor using a mobile dipping bath and a diazinon based dip. Sheep dipping contractors have greater potential for exposure to organophosphates in sheep dip than other sheep farmers because of the high numbers of sheep they dip.

The personal sampling apparatus was placed in the breathing zone of the operative and consisted of a two stage collection device (OVS 2, SKC Ltd.). The first part consisted of a filter (GFA) designed to collect any particulate or aerosol present and the second part contained an adsorbent (XAD-2) to collect any vapour present. The two stages were in series and connected to a calibrated pump which pulled a known volume of air through the collection device. If during the process some of the 'aerosol' on the filter were to evaporate it would be collected on the back-up adsorbent tube. In addition to the personal samplers, static samplers were placed in the vicinity to determine general environmental levels of diazinon. Details are given below for the analytical procedure used to quantify the amount of diazinon on the filters and adsorbent.

2.1.3 Other groups studied for urinary dialkyl phosphates. Urinary metabolite data from the studies of workers occupationally exposed to sheep dip described here have been compared with the urinary results from other occupational groups. Some of these groups also had potential for exposure to organophosphorus pesticides at work whilst other groups were used as control populations whose work was not expected to bring them into contact with pesticides. The populations were defined as:

(a) Foundry workers - a control group of workers with no known occupational exposure to pesticides with repeat samples collected over the course of a week;

(b) Office workers - a larger control group with no known occupational exposure to pesticides with one urine sample collected from each volunteer;

(c) Sheep dippers - workers involved in dipping sheep with diazinon or chlorfenvinphos;

(d) Agricultural workers - farm workers using organophosphate pesticides to spray crops;

(e) Formulators - workers involved in formulating pesticides.

Urine samples from exposed groups were subdivided into:

(a) ' Pre' exposure - urines from workers just before the use of organophosphorus pesticides that might be metabolised to the dialkyl phosphates of interest;

(b) 'Post' exposure - samples collected from workers within four days of the use of these organophosphorus pesticides.

In addition, 24 hr urine samples were analysed from a woman admitted to hospital following a suicide attempt with chlorpyrifos. Urine samples were collected from the time of admission until she was released 11 days later. All these data have been used to help put the results from the current studies into context.

2.2 Analytical Procedures

2.2.1 Urinary Dialkyl Phosphates. Urinary dialkyl phosphates were analysed by high resolution gas chromatography as their pentafluorobenzyl ester derivatives using a previously published method[7]. Briefly, calibration curves (range 0 - 3 μmole litre^{-1}) were prepared using urine from laboratory personnel with no known exposure to organophosphates. The control urine (1 ml) was spiked with the six metabolites of interest - DMP, DMPT, DMPDT, DEP, DEPT and DEPDT.

Quality control samples were prepared using urine from workers not exposed to organophosphorus pesticides which was spiked with the six metabolites at a nominal 1 μmole litre^{-1} . These samples were then frozen until required. Analytical performance was assessed by inclusion of quality control samples in each run. These were added after the calibration curve, after each set of five urine samples (analysed in duplicate) and after the final sample.

Dibutylphosphate (250 μl, 3 μmole litre^{-1} solution) was added as internal standard to all calibration, unknown and quality control urine samples (1 ml). These were thoroughly mixed with acetonitrile (8 ml) and centrifuged for 10 min. The supernatant was transferred to a clean tube and anti-bumping granules added. The tubes were evaporated to dryness under a stream of oxygen free nitrogen at a temperature of 90°C. Once dry a further aliquot of acetonitrile (4 ml) was added to each tube and the drying procedure repeated. Anhydrous potassium carbonate (c.50 mg) was added to each tube along with acetonitrile (0.5 ml) and 2,3,4,5,6-pentafluorobenzyl bromide (25 μl) and thoroughly mixed. Tubes were then capped and heated overnight (12 - 16 hr) at 50°C. After cooling the supernatant was transferred to an autosampler vial, sealed and subjected to high resolution capillary gas chromatography with flame photometric detection.

Gas chromatographic analysis used an HP 5890 Series II gas chromatograph fitted with an HP 7673B autosampler, split-splitless injector and flame photometric detector. Injections (1 μl) were made with an injection temperature of 280°C into a BP10 25 m x 0.33 mm i.d. (0.5 μm thick film) capilllary column with a helium carrier gas flow rate of 1 ml min^{-1}. The initial oven temperature of 140°C was held for 1 min, then increased at 8°C min^{-1} to 280°C and held for 2 min. The flame photometric detector was operated in the phosphorus mode and gas flow rates set for maximum sensitivity and specificity for the dialkyl phosphate derivatives.

2.2.2 Environmental samples. Filters (GFA) and sorbent (XAD-2) were analysed using modifications of NIOSH guideline methods[8]. Filters and sorbent were extracted separately with toluene (2 ml), left to stand for 2 hr and an aliquot of the solvent transferred to separate GC vials, capped and analysed.

All extracted samples were analysed by high resolution gas chromatography with nitrogen phosphorus detection. Gas chromatographic analysis used an HP 5880 gas chromatograph fitted with an HP 7673B autosampler, split-splitless injector and a nitrogen phosphorus detector. Injections (1 μl) were made with an injection temperature of 250°C into a PA5-1701 capilllary column with a helium carrier gas flow rate of 0.2 ml min^{-1}. The initial oven temperature of 60°C was held for 1 min, then increased at 20°C min^{-1} to 160°C, then at 5°C min^{-1} to 210°C and held for 1 min. Oven temperature was then increased to 290°C to remove unwanted compounds from the column.

The system was calibrated with diazinon standards over the range 0.05 - 10.0 μg ml^{-1}. Peak areas were measured and plotted against concentration of standards and used to quantify diazinon present in the extracts. Calibration curves were linear over the calibration range (r^2=0.9999).

2.2.3 Fluorescence measurements. Fluorescence measurements were made in the field using a mobile laboratory. After the operatives involved in dipping had finished preparing the dip bath by diluting dip concentrate to the required strength fluorescent dye was added to the bath. Dye was added by laboratory staff to minimise the risk of false contamination of farm workers. The ratio of concentration of dye to concentration of active ingredient in the dip was noted. The workers continued with their dipping activities until all sheep had been dipped. If during dipping additional concentrate was added to the dip bath to maintain a suitable concentration of active ingredient the appropriate amount of dye was also added.

After dipping was complete each operative was asked to stand inside a shell of ultra-violet lights which makes the dye fluoresce where it has landed on the skin. The intensity of fluorescence was measured using a monochromatic video camera. The subjects were photographed in various poses to cover the entire body. Notes were made detailing the location of the fluorescence, which appeared as a blue glow, to prevent subsequent confusion with natural skin fluorescence.

The fluorescence system was calibrated using a member of laboratory staff. A known area of the volunteers' skin was outlined with an indelible marker. A series of known volumes of known concentrations of dye were spread evenly over this area to produce known surface concentrations which were photographed when dry. The average intensities of the glow in each image were measured, a calibration curve relating intensity to concentration was produced and subsequently programmed into an image processor. Farmers were individually spot calibrated for fluorescent dye intensity to allow for system compensation of normal skin fluorescence for each subject.

Subject images were displayed on a TV monitor. Each area within the image that showed fluorescence was isolated, the background fluorescence of nearby uncontaminated skin estimated and the difference between the two determined. This was then used to estimate the extent of skin contamination with diluted dip.

3 RESULTS

3.1 Urinary dialkyl phosphates

Results for urinary dialkyl phosphates are reported following correction for urinary creatinine concentration. Creatinine is a waste product of muscle metabolism produced at a fairly constant rate and virtually completely cleared from blood during its passage through kidneys. Creatinine correction is frequently used to compensate for differences in urine volume which would affect absolute metabolite concentrations. To simplify interpretation of the results and to overcome the effect on differences in relative activities of the biotransformation pathways leading to production of these metabolites urinary metabolite values are expressed as the sum of DEP and DEPT. Results are reported as the mean and range of values found and the 90% value i.e. the sample result below which more than 90% of results of a group are found.

Urinary total 'ethyl' phosphate metabolite levels for each sampling period (pre dipping, post dipping or next day) for all subjects are reported in Table 1. Results from urine samples obtained under similar conditions from other workers including other sheep dippers are given in Table 2.

In addition, 90% values for urine samples from the unexposed and potentially exposed occupational groups are compared with metabolite levels detected in urine from a women who attempted suicide using chlorpyrifos (Figure 2). Chlorpyrifos, like diazinon, is metabolised to DEP and DEPT.

Table 1 *Total DEP and DEPT found in urines from all sheep dippers in the survey.*

Urine collection time	Mean and (range) total urinary 'ethyl' phosphate metabolites/ nmol/mmol creatinine	90% value/ nmol/mmol creatinine
Pre dipping	7 (0 - 65)	14
Post dipping	11 (0 - 95)	28
Next day	25 (0 - 154)	48

Table 2 *Urinary dialkyl phosphate data from various occupational groups and workers with no known occupational exposure to organophosporus pesticides*

Occupational Groups (sampling time)	Mean and (range) total urinary 'ethyl' phosphate metabolites/ nmol/mmol creatinine	90% of results less than	Comments
Foundry workers	1.1 (0 - 24)	5	261 urine samples from 25 individuals with no known <u>occupational</u> exposure to organophosphorus pesticides
Office workers	1.6 (0 - 32)	6	106 samples from 106 individuals with no known <u>occupational</u> exposure to organophosphorus pesticides
Sheep dippers ('pre' dipping)	5.6 (0 - 162)	14	159 samples from 159 individuals (obtained from other studies of sheep dippers)
Sheep dippers ('post' dipping)	15 (0 - 189)	40	337 urine samples from 167 individuals (obtained from other studies of sheep dippers)
Agricultural workers ('pre' exposure)	1.3 (0 - 17)	5	35 urine samples from 35 individuals
Agricultural workers ('post' exposure)	10.3 (0 - 159)	27	59 urine samples from 35 individuals
Formulators	42.8 (0 - 479)	72	88 urine samples from 10 individuals

Dialkyl phosphate metabolites that could be derived from either chlorfenvinphos (DEP) or diazinon (DEP and DEPT) were found in 15 pre dipping urine samples. However, the levels were similar to those seen in foundry and office workers who were not occupationally exposed to organophosphorus compounds. Although attempts were made to study farmers on the first day of dipping of the new season some farmers may have been exposed to other organophosphorus pesticides such as those used for treating crops or other insect pests of animals. Additionally the metabolites seen in workers with no known occupational exposure to organophosphorus pesticides suggests that absorption of organophosphorus pesticides following non-occupational exposure, possibly from dietary sources or domestic use of products containing these compounds, may also occur.

Metabolites seen in post dipping urines, either on the same day or the day after dipping, showed an increase in the mean, range and 90% values when compared with 'pre' dipping values. Generally, the highest metabolite levels were seen in the next day urines although there were some exceptions. A significant difference was found between adjusted next day urine samples from those operators who handled concentrated dip and those that did not ($p<0.01$). This suggests that skin contamination with dip concentrate is an important source of exposure.

Comparison of post dipping metabolite levels from workers in the study described here (Table 1) with results from other occupational groups (Table 2) suggests that the extent of absorption of organophosphorus pesticides present in sheep dip observed in this study was of a similar magnitude to that found with other sheep dippers or other farm workers studied. The data in Table 2 suggests that absorption of these pesticides in all agricultural workers is lower than that for formulators.

Metabolite levels in urine from workers occupationally exposed to organophosphorus pesticides are substantially below those detected following an acute poisoning (attempted suicide) with chlorpyrifos (Figure 2). In this case the maximum urine concentration of total 'ethyl' metabolites was 185,000 nmol/mmol creatinine on the day after the 'exposure'. When the patient was discharged from hospital 10 days later (with cholinesterase returning to normal) the urine metabolite concentration was 15,000 nmol/mmol creatinine. These metabolite levels are almost 1000 times higher than those seen to date in any urine samples received from sheep dippers.

3.2 Air sampling

Results for airborne concentrations of the organophosphorus pesticide diazinon detected during sheep dipping are given in Table 3. The results are expressed as the amount detected over the actual period of dipping. If an 8 hr time weighted average (TWA) was calculated the results would be even lower. In the UK the 8 hr TWA Occupational Exposure Standard (OES) for diazinon is 100 μg m^{-3} which is approximately 3 times the actual value obtained over the sampling period (about 3 hr)[9].

Levels of diazinon detected in static samplers placed near the mobile dipping bath were also low being approximately a tenth of the 8 hr TWA value for static sampler 1 which was approximately 0.8 m from the dip bath and nearer to the dipping operation than sampler 2.

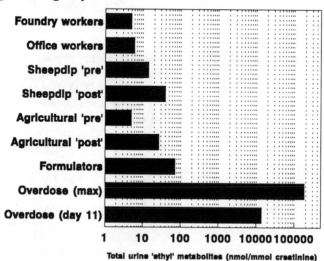

Figure 2 *Comparison of urine metabolite values of occupational groups with overdose case*

Table 3 *Results from personal and static sampling for airborne levels of organophosphorus pesticides during sheep dipping.*

	Airborne concentration/$\mu g\ m^{-3}$	
Sampler type	GFA filter	XAD-2 sorbent
Personal	14.3	21.1
Static 1	6.6	5.5
Static 2	2.3	Not detected

3.3 Fluorescence monitoring

Results from fluorescence monitoring of dermal contamination for each subject by body region are summarised in Table 4. Dermal doses of diazinon were calculated from dermal measurements of fluorescence after background subtraction using the theoretical ratios of diazinon to dye added to the dipping bath. This makes no allowance for the possibility that compared with the dye diazinon was preferentially 'stripped out' of the dipping solution.

Table 4 *Summary of data from fluorescent monitoring of workers exposed to dilute sheep dip during dipping operations.*

Subject Number	Dermal diazinon dose/μg								Notes
	Front	Back	Head	L. Arm	R. Arm	L. Leg	R. Leg	Total	
1			3	78	36			120	2d
2	5		30	110	60			210	2d
3	75	0.7	70	220	50	240^z	60^z	720	2d, z
4	0.15			0.3	0.25			.07	
5	.01			3.3	0.5			3.8	p
6	0.1		1	2.4	12			>15	p
7			8.8	>0.12	>3			>12	p
8			12	460	230			700	
9			0.2	60	90			150	2d, mud
10	48	16	15	>80	>60		85	>310	2d
11	1.2		8.4	113	290			410	z
12			12	29	9^z			50	
13			10	70	70			150	
14			0.5	0.8				1.3	
15			5	80^z	55			140	z
16			0.7	14.6	8			23	

Notes
2d = 2-dimensional measurements only. Allowance has been made for skin curvature.
z = high background zeroes
p = poor quality images
Mud = hands covered with mud. The results are an underestimate.

Contamination of workers measured was low with 720 and 700 μg being the highest values obtained. The results are likely to be underestimates, because mud on hands masked some of the fluorescence. On several occasions a fault in the fluorescence monitor lighting system occured, which resulted in underestimation of skin surface areas, an allowance was made for curvature based on previous experience. These results are marked by the prefix 2d in Table 4. Some subjects had high background fluorescence (prefix z in Table 4) which may have increased the errors of measurement; to overcome this background measurements were measured pre-exposure. However, this was not possible on the legs of subject 3 and hands of subjects 11 and 15. Poor quality images were obtained with some subjects (prefix p in Table 4) which also reduced the accuracy of fluorescence measured.

4 DISCUSSION

The studies reported here show that occupational exposure to organophosphorus pesticides can occur during sheep dipping operations. Three analytical techniques were used to investigate this occupational hygiene problem. Biological monitoring was used to investigate the extent of absorption of the pesticides present in sheep dip. The data obtained, when combined with observations made by occupational hygienists present during dippping, suggests that handling of concentrate is an important source of exposure of workers to these organophosphorus pesticides. During preparation of dip baths for use concentrate is usually diluted 1 part in 1500. Therefore, exposure to 100 μl of concentrate would be equivalent to exposure to 150 ml of dilute dip solution.

Although exposure to concentrate may be a significant source of contamination, the data from fluorescence monitoring studies show that exposure to dilute dip can occur. The qualitative and quantitative results indicate that contamination was generally highest on the face, hands and arms. Estimates of the amount of pesticide retained on the skin suggest that dermal contamination was less than 1 mg in total. However, the fluorescence measurements are probably underestimates.

No correlation was found between urine metabolite levels and estimates of dermal exposure from fluoresence measurements. As metabolite concentration is related to the total dose absorbed following exposure to both concentrate and dilute dip and fluorescence measurements estimate exposure to dilute dip only, a lack of correlation may be expected. In addition, the rate of dermal absorption of chemicals is proportional to the concentration on the skin. Therefore, organophosphate pesticides in the concentrate will be more rapidly absorbed into the body when compared with that present in the dilute dip.

Measurement of airborne organophosphorus pesticides, present either as an aerosol or a vapour, indicates that duing sheep dipping operations inhalation is a minor route of exposure. The levels of diazinon detected over a 3 hr period were less than a third of the UK 8 hr TWA occupational exposure limit for diazinon of 100 μgm^{-3}. These data support the findings of Niven et al[7] who reported that in a pilot study airborne levels of diazinon were low and did not exceed the detection limit of the assay which was equivalent to exposure to diazinon of 10 μgm^{-3} for a 4 hr period.

The studies reported here were not designed to look at the health effects of exposure to organophosphorus. However, urinary dialkyl phosphate data obtained following an unsucccessful suicide attempt with the organophosphate pesticide chlorpyrifos shows that during gross acute exposure total DEP and DEPT levels were over 1000 times those seen in urine from sheep dippers. An interesting observation from this overdose case was that on the eleventh day after admission, when the patient was well enough to be discharged from hospital, the patient's total DEP and DETP excretion was still approximately 100 times that seen in urine samples from sheep dippers. However, at such excessive doses some of the metabolic processes involved in detoxifying these pesticides may be saturated and lead to different metabolite profiles to those seen at the lower exposure levels found during occupational use of organophosphorus sheep dips.

The different techniques used in this occupational hygiene study have investigated several different aspects of exposure to sheep dip. Proper assessment of workplace practices is important and serves to provide a descriptive account of the processes involved. Quantitative estimates of exposure provided by biological, environmental and fluorescent monitoring are important for assessing the toxicological significance of hygiene observations. These complementary approaches are important and provide data to allow regulatory authorities to produce health and safety guidance based on scientific studies.

5 ACKNOWLEDGEMENTS

The authors would like to thank the West Midlands Poisons Unit, Birmingham for their assistance with these studies.

6 REFERENCES

1. Ministry of Agriculture, Fisheries and Food, 'Agriculture in the UK', Her Majesty's Stationary Office, 1992.
2. 'Organophosphorus Pesticides. Criteria (Dose/Effect Relationships) for Organophosphorus Pesticides', Editor. R. Derache, Pergamon Press, 1988.
3. 'Casarrat and Doull's Toxicology. The Basic Science of Poisons', 4th Edition, Editors M. O. Amdur, J. Doull and C. D. Klaassen, Pergamon Press, 1991.
4. 'Clinical and Experimental Toxicology of Organophosphates and Carbamates', Editors B. Ballantyne and T. C. Marrs, Butterworth-Heineman, 1992.
5. E. P. Savage, T. J. Keefe, L. M. Mounce, R. K. Heaton, J. A. Lewis and P. J. Burcar, *Archives of Environmental Health*, 1988, **43**, 38.
6. K. J. M. Niven, A. J. Scott, S. Hagen, E. R. Waclawski, M. Lovett, B. Cherrie, P. L. Bodsworth, A. Robertson, A. Elder, J. Cocker, B. P. Nutley amd M. Roff, 'Occupational hygiene assessment of sheep dipping practices and processes', 1993, TM/93/03. Institute of Occupational Medicine Technical Memorandum Series.
7. B. P. Nutley and J. Cocker, *Pesticide Science*, 1993, **38**, 315.
8. National Institute of Occupational Safety and Health 'Manual of Analytical Methods', 4th Edition, 1992.
9. Health and Safety Executive, 'EH40/95 Occupational Exposure Limits 1995', HSE Books, 1995.

Application Technology

G. A. Matthews

IPARC, DEPARTMENT OF BIOLOGY, IMPERIAL COLLEGE OF SCIENCE, TECHNOLOGY AND
MEDICINE, SILWOOD PARK, ASCOT, BERKSHIRE SL5 7PY, UK

1. INTRODUCTION

The public are concerned about pesticide residues in the environment, especially in
water supplies and food, while the user is more concerned about avoiding being
contaminated while applying the pesticides, and minimising drift to neighbouring areas.
Environmentalists are also concerned about possible effects of pesticide drift to Sites of
Special Scientific Interest (SSSIs) as well as the countryside [1]. Despite these concerns
there has been relatively little change in the basic way in which pesticides are applied [2].
Apart from a few specialised application techniques, a pesticide is formulated to be
mixed with water and the diluted mixture pumped through one or more hydraulic
nozzles. Sprays were applied using very large volumes of water to thoroughly wet
surfaces with diluted pesticide, but for various reasons including the logistics of
transporting the water, most applications are now with less than 250 litres per hectare.

With the Food and Environmental Protection legislation and concerns of the Health and
Safety Executive, some changes in choice and design of equipment and packaging have
been initiated. Some of these changes are discussed with particular reference to the
consequences for greater efficiency in application and less wastage to non-target areas.
Much of this paper refers to field applications, but in terms of concern about residue
analysis, application techniques used in the more intensive horticultural industry and to
treat produce in stores are also discussed, as the residues in these situations are directly
on marketable produce.

2. SPRAY DRIFT

The dominant form of pesticide spray application in the United Kingdom is with a
tractor-mounted or trailed sprayer. In arable farming, a series of hydraulic nozzles are
mounted on a horizontal boom, typically about 50cm above the crop or ground. In tree
and other crops, the nozzles are usually mounted in an air-stream to project the spray
droplets to the crop canopy. Hydraulic pressure nozzles produce a wide range of droplet
sizes, with the proportion of small droplets, less than 100µm in diameter, increasing at
higher operating pressures or with a smaller nozzle orifice. The subsequent movement
of these droplets is affected by their size, and air movement between the nozzle and the
crop. Apart from natural air movement, turbulence is created by the forward movement

of the sprayer; thus even at 2m/s forward speed, droplets less than 100μm in diameter are liable to become airborne, and more prone to downwind transport. Once airborne droplets can drift and be transported beyond the field boundary. Measuring the amount of drift is very difficult due to problems of sampling very small airborne droplets [3].

The potential for spray drift is increased in hot dry weather, with convective turbulence, when droplets will decrease in size due to evaporation of the water carrier. The lifetime of small water droplets (<100μm) is generally less than a minute under normal conditions, so after the carrier has evaporated, only an aerosol particle of active ingredient and other involatile components of the spray remain. To minimise adverse effect of drift, an unsprayed barrier zone is left next to hedgerows and other wildlife habitats, or spraying is confined to periods when the wind is blowing away from sensitive areas. Several approaches to minimising the potential for drift have been proposed. These involve changes in nozzle selection and operating pressure, providing air assistance to project the spray into a crop canopy, shielding the spray from the wind and adding an electrostatic charge to the spray droplets.

2.1 Nozzle Selection.

Spray nozzles are now classified according to the droplet spectrum produced. A series of reference standard fan nozzles have been designated to indicate the spectrum produced by fine, medium and coarse sprays [4]. One nozzle was also selected to demarcate the division between the fine and very fine category. Subsequent discussions in Europe are extending the system with extra reference nozzles for the boundaries of the other categories. As the numerical values may differ between measuring systems this procedure allows different laboratories to use their own measuring equipment to compare data from the reference nozzles with any other nozzles being examined. Typical laser light diffraction equipment can be used to measure the spectrum at a standard distance from the nozzle [5]. Wind tunnel studies have generally confirmed the classification system [6].

Nozzles defined as producing a very fine spray are not allowed to be used in arable crops, due to the high proportion of 'driftable' spray. However, with all hydraulic nozzles currently used, there is a proportion of the spray as small droplets liable to drift, even when coarse sprays are applied using nozzles with a larger orifice. The volume of spray lost as drift may be only a small fraction of the total spray, whereas a higher proportion of the volume in large droplets can be lost at the time of the application directly to the soil, or later due to removal of surface residues on foliage by rain. Soil contamination may adversely affect soil-inhabiting beneficial species, and depending on the physical properties of the pesticide and extent of degradation by micro-organisms, these residues can contaminate ground water or local waterways if exposed to surface 'run-off'. The greater concern for the perceived losses due to drift has greater use of a range of pre-orifice nozzles, which for a given output and operating pressure will provide a coarser spray, less prone to drift [7,8]. However with a larger average droplet size, spray distribution within a crop canopy and coverage of individual leaves may be adversely affected. Furthermore there is greater risk of more spray being deposited on the soil under the crop.

Alternative nozzles include a twin-fluid nozzle in which compressed air is delivered into a modified deflector nozzle [9,10]. This nozzle enables lower volumes to be applied without having a very small orifice. Provided the air and liquid pressures are carefully controlled, a coarser spray is produced with some aerated droplets and an air flow to assist delivery of the spray [11,12]. An adaptation of a fan nozzle eliminates the need for using an air compressor, as air is sucked into the nozzle by a venturi through two small holes in the side of the nozzle.

Rotary nozzles produce a narrower droplet spectrum that can be adjusted by control of the speed of rotation and flow rate. Their main use in the United Kingdom has been with hand-held equipment, particularly with large droplets (c. 250μm or larger) to apply herbicides without drift. Research with tractor-mounted boom sprayers was extensively carried out with reduced volumes of application (40-60l/ha) [13]. Despite offering one of the best methods of controlling the droplet spectrum, their use has not been widely adopted for various reasons, including difficulties in the reliability of drive systems, and lack of spray penetration of some crop canopies. These rotary nozzles have been more widely used to apply insecticides in the tropics [14].

2.1.1 Pressure control   On the tractor equipment there is usually a pressure control, valve so that the spray is delivered to the boom at a constant pressure. However, depending on the boom length and delivery hose capacity, there can be a decrease in pressure along a boom. In some designs of sprayer the pressure is adjusted in relation to forward speed to maintain a constant volume application rate. In the United Kingdom, most tractor sprayers are operated at speeds of approximately 7km/h, but in other countries, speeds of up to 20km/h are known to be used. With wide fluctuations in forward speed, there are variations in nozzle pressure that can have a significant effect on the proportion of spray droplets that are liable to drift.

Most boom sprayers are fitted with a diaphragm check valve, which prevents spray liquid dripping from the boom when the operator stops at the edge of a field. However, a new design of valve referred to as a spray management valve not only stops dripping in the same way as a diaphragm check valve, but also regulates the output pressure at the nozzle.   Thus for herbicide application a 1 bar pressure can be maintained. The spray management valve has been introduced initially on knapsack sprayers where the pressure is more variable due to the influence of the operator on pumping rate [15]. The use of a spray pressure control at each nozzle on a tractor boom would reduce the risk of drift, but there would be less flexibility in terms of forward speed.

2.2 Air assisted sprayers.

Another approach has been the development of air-assisted booms in which air is delivered from a fan to a sleeve mounted alongside the spray boom so that a jet or curtain of air entrains the droplets and projects them into the crop canopy [16]. These were designed to provide better penetration of the crop canopy and control pests and diseases in the lower canopy [17], but when there is sufficient foliage to filter the droplets from the airstream, their use also reduces downwind drift.   The airsleeve should not be used if there are small plants, or the aim is a soil surface treatment, as the air is deflected back and droplets are still liable to drift.

Air-assisted sprayers, commonly referred to as mistblowers, have been used more extensively for treating orchard crops. Due to the height of traditional tree crops, such as apples, air was used to project spray to the top of the canopy, which could result in significant spray drift from above the trees. The trend to using dwarf root stocks and other changes in the planting of orchards has enabled development of other equipment. Some sprayers now use cross-flow fans close to the crop canopy, while other manufacturers have designed 'tunnel sprayers', in which a mobile canopy protects the tree from a crosswind during application. Spray which passes through the canopy is impacted on the tunnel and recycled.

Air assistance is particularly needed in glasshouses and stores to circulate very small spray droplets, generally <50μm in diameter, needed to percolate through the airspace between vegetation or crates of produce. These treatments, often referred to as fogs, contain a high proportion of inhalable and respirable particles, so treatment from outside through ducting or by remote control is preferred when the building is unoccupied. Problems of increased residues have occurred, presumably where the small particles have sedimented eventually on horizontal surfaces in still air.

2.3 Electrostatic sprayers.

The charging of spray droplets electrostatically will increase deposition on the nearest earthed object. In field crops, the top of plants received more spray deposition, but penetration of a crop canopy was often inadequate to control some pests [18]. Drift of highly charged small droplets was significantly reduced, unless pockets of convective air turbulence carried these small droplets above the crop [19]. At present a suitable technique for large arable farms has not been commercially developed, whereas in the tropics a small hand-held electrostatic sprayer has been used for spraying cotton with a pre-packaged ultra-low volume oil-based formulation, provided in a restricted range of insecticides [20,21]. One type of electrostatic sprayer was designed specifically for fungicide treatment of potatoes [22].

2.4 Pesticide transfer

The main concern for the user is contamination with the concentrated pesticide during its transfer from the container to the sprayer. Recommendations have been made for extra personal protective clothing, such as an apron, face shield and gloves, to minimise contamination of the operator's face and overalls due to splashes incurred while opening a container or pouring the concentrate into a mixing unit or direct to the sprayer. However, the use of protective clothing is regarded as a last resort, and the HSE has encouraged the development of engineering controls to minimise exposure during pesticide transfer. Several systems have been or are being developed.

2.4.1 Packaging systems   Most of the new systems of transfer depend on greater uniformity of the original packaging, so the agrochemical industry has agreed to standardise manually handled pesticide containers. All containers with a capacity > 1 litre now have an opening of 63mm diameter, that allows the contents to be poured if necessary with minimal risk of splashing, or coupled directly to a transfer system. Such containers should be rinsed immediately and the washings added to the sprayer tank, to avoid contaminated containers polluting the environment.   The use of a closed transfer system is preferred, so some products are now available in 10 litre or larger refillable containers, but these are returned un-rinsed to the supplier [23,24].   Already in some countries, such as Canada a surcharge on each container to pay for disposal has accelerated the adoption of refillable containers.   Dry products, such as dispersible grains, can be packed in pre-measured quantities in water-soluble sachets, based on polyvinylalcohol films that can be placed directly into the sprayer or a low level induction bowl [25].

2.4.2 Low level induction bowls   On the old type of spray equipment, the operator would climb on the sprayer to pour concentrate through an opening on the top of the tank. This procedure is now regarded as hazardous. Apart from the problem of lifting a large heavy container, there is a risk of falling with a hazardous chemical. There has now been rapid adoption of fitting a low-level induction bowl or hopper to sprayers, mounted between 0.5 and 1m above ground level. It should be equipped with a rinsing device that will transfer any pesticide placed in it to the main sprayer tank and leave no more than a residue of 1ml. of the original pesticide in the bowl [26].

2.4.3 Closed chemical transfer systems (CCTS)   Instead of pouring pesticide into the sprayer or mixing unit, the pesticide container is directly coupled to an apparatus that will allow   the entire contents of the container to be transferred   to the sprayer or a measured dosage, in which case any excess pesticide has to be returned to the container before separation from the system [27]. The measurement system used for non-returnable containers must provide a method of rinsing so that all surfaces are cleaned to reduce any residue to less than 0.01% of the volume of the item rinsed.   Similarly   the measuring system shall be rinsed after the separation of returnable containers, and the rinsate added to the sprayer tank. The apparatus must function at or below atmospheric pressure to minimise uncontrolled escape due to component failure or damage. When couplings are broken, no more than 1g of liquid should escape.   Similar apparatus may be used for transfer of dry products and gels. The controls, positioned no higher than 1 m above the ground, must be easy to operate when using gloved hands. The apparatus must also be designed to ensure there can be no siphoning of pesticide from the sprayer.

2.2.4 Injection systems.   For a long time it has been considered desirable to have a system whereby the pesticide concentrate is only mixed with water as it is applied [28]. The need to dispose of any surplus dilute spray is eliminated and the tank is not contaminated with pesticide. Several systems of injecting a pesticide into the water carrier have been developed, but many of these are quite complex and expensive, so few are used. Nau and Raffel [29] describe a system with a precision pump driven by the water flow so that the amount of pesticide injected is proportional to the volume of water displaced. Frost used water pressure [30] to activate a piston that metered pesticide into the spray line to the boom. Subsequent research has focussed on the prospects of an injection system in conjunction with localised 'patch' spraying of weeds [31,32].

Another injection system developed initially for knapsack sprayers uses a flexible bag inside a container; water under pressure enters the container and as the bag is squeezed, pesticide is fed through a metering orifice into the spray line [33]. The metering rate and concentration of the pesticide are governed by the metering orifice fitted by the supplier, thus eliminating the risk of the operator using the wrong spray concentration. The dispenser is designed to be returned to the supplier for refilling, but so far no supplier has adopted this technology, in part due to the costs inherent in recycling large numbers of small containers.

3.1 Other methods of application

The greatest risk of contamination during application is inevitably with pedestrian equipment, where spray is released close to the operator. The problem is less when large droplets are applied close to the ground, in contrast to treatment of tree crops with a finer spray. As indicated earlier, downwind movement of droplets is exploited with some hand-held equipment with rotary nozzles, but operator contamination with other equipment, including knapsack sprayers would be reduced by keeping nozzles downwind of the body. The greatest problem is with treatments inside buildings especially where space sprays (fogs and mists) are applied in stores and glasshouses. In some situations a fog can be introduced from outside or automatically at a pre-set time to eliminate exposure of the operator during treatment.

Operator contamination is less with granule application, especially when packaging allows direct transfer of the pesticide to the hopper. Other specialised application techniques that allow more precise placement of the pesticide include seed treatment and the use of weed wipers.

Aerial spraying is still needed in forestry and certain other situations where access with ground equipment is restricted. However the demand in arable crop situations has declined as growers now use 'tramlines' to allow access with tractor sprayers throughout the season.

4.1 DISCUSSION

Highly persistent pesticides, such as the organochlorine insecticides have been detected in all parts of the world, even in areas remote from large scale applications. This is no doubt due to the movement of aerosol and sub-micron sized particles in the atmosphere over considerable distances from where they were released. In particular the release of water-based sprays through hydraulic nozzles produces a range of droplet sizes, and although the volume of spray with droplets <100µm in diameter is a small fraction of the total spray, the number of droplets can be extremely large. As evaporation results in a decrease in their volume, sampling and measurement of the extent of drift is difficult, yet the effects of some pesticides are very easily detected, for example the effects of certain insecticides on bees and herbicides on sensitive plants [34]. The Spray Classification system is a means of improving the selection of hydraulic nozzles for a particular application, but there is a reluctance to change to other application techniques. Hydraulic nozzles are relatively inexpensive, and are easily interchanged to provide versatility for different pesticides, crops and farms. In contrast, the alternatives so far developed, such as the rotary atomisers, are more specialised, although they have a role in certain situations.

A system for classifying equipment in relation to the potential hazard to the operator and the environment has been proposed [35]. The extent to which operator and environmental contamination can be reduced is dependent upon the level of training, observance of instructions and use of safety systems, including wearing personal protective equipment. Operator training is an important feature in the use of pesticides in the United Kingdom. Methods of operating equipment need to be kept as simple as possible with easy and reliable adjustment, such as nozzle replacement and pressure regulation, to counteract effects due to wear following normal use of the equipment. Provision of a clean water tank for washing is a new requirement to enhance the safety of the operator. In some countries this water tank must have sufficient capacity to allow washing out of the sprayer tank while the equipment is still in the field so that the washings can be sprayed on the last sector of the treated field.

Most emphasis in recent years has been on changes in formulations to suspension concentrates and dispersible granules which reduce the need for organic solvents, such as xylene, in typical emulsifiable concentrates. These changes together with new packaging standards have been aimed principally at increased safety of use. Their adoption has been stimulated by legislation, that overcomes reluctance to increase costs. Rinsates from containers can be used as diluent in sprays, but the washings of equipment and disposal of surplus diluted spray require special treatment to minimise environmental pollution. Equipment has been designed to facilitate removal of pesticides from water used in agricultural spraying operations [36].

References

1. A. S. Cooke (Ed) "The Environmental Effects of Pesticide Drift." English Nature, Peterborough, 1993.
2. G. A. Matthews, "Pesticide Application Methods." 2nd Edition, Longman, Harlow, 1992.
3. P. C. H. Miller, In "Application Technology for Crop Protection", G. A. Matthews and E. C. Hislop (Eds.) CABI Wallingford. 101, 1993.
4. S. J. Doble, G. A. Matthews, I. Rutherford, and E. S. E. Southcombe, *Proc. British Crop Protection Conference - Weeds*, 1983, 1125.
5. C. S. Parkin, In "Application Technology for Crop Protection", G. A. Matthews and E. C. Hislop (Eds) CABI Wallingford. 57, 1993.
6. P. C. H. Miller, E. C. Hislop, C. S. Parkin, G. A. Matthews, and A. J. Gilbert, *ANPP-BCPC 2nd International Symposium on Pesticide Application Techniques.* Strasbourg, 1993, 109.
7. G. S. Barnett and G. A. Matthews, *Int. Pest Control.*, 1992, **34**, 81.
8. J. A. Castell, *ANPP-BCPC 2nd International Symposium on Pesticide Application Techniques.* Strasbourg, 1993, 227.
9. C. Cowell and A. Lavers, *Aspects of Applied Biology*, 1987, **14**, 35.
10. P. C. H. Miller, C .R. Juck, A. J. Gilbert and G. J. Bell, *BCPC Monograph*, 1991, **46**, 97.

11. N. M. Western, E. C. Hislop, P. J. Herrington and E. I. Jones, *Proc. Brighton Crop Protection Conference - Weeds,* 1989, 641.

12. B. W. Young, *BCPC Monograph,* 1991, **46**, 77.

13. W. A. Taylor and C. R. Merritt, *Proc. 8th British Insecticide and Fungicide Conference,* 1975, 161.

14. J. S. Clayton, T. E. Bals and G. S. Povey, *ANPP-BCPC 2nd International Symposium on Pesticide Application Techniques.* Strasbourg, 1993,199

15. G. A. Matthews and E. W. Thornhill, *Proc. Brighton Crop Protection Council Conference - Weeds,* 1993,1171.

16. W. A. Taylor and P. G. Andersen, *BCPC Monograph,* 1991, **46**, 125.

17. E. Nordbo and W. A. Taylor, *BCPC Monograph,* 1991, **46**, 113.

18. G. R. Cayley, P. R. Etheridge, D. C. Griffiths, F. T. Phillips, B. J. Pye and G. C. Scott, *Ann. appl. Biol.,* 1984, **105**, 279.

19. N. M. Western and E. C. Hislop, *BCPC Monograph,* 1991, **46**, 69.

20. R. A. Coffee, *Proc. British Crop Protection Conference - Pests and Diseases,*1979, **3,** 777.

21. G. A. Matthews, *Crop Protection,* 1989, **8,** 3.

22. G. R. Cayley, G. A. Hide, J. Lewthwaite, B. J. Pye and P. J. Vojvodic, *Potato Research,* 1987, **30,** 310.

23. T. H. Robinson and J. P. Cowland, *Proc. Brighton Crop Protection Council Conference - Pests and Diseases,* 1994, 985.

24. J. M. Ogilvy and K. L. Tyler, *Proc. Brighton Crop Protection Conference - Pests and Diseases,* 1994, 991.

25. J. Hartmann, *ANPP-BCPC 2nd International Symposium on Pesticide Application Techniques.* Strasbourg, 1993, 47.

26. A. J. Gilbert and C. R. Glass, *Proc. Brighton Crop Protection Council Conference - Weeds,* 1991, 693.

27. R. H. Garnett, *ANPP-BCPC 2nd International Symposium on Pesticide Application Techniques.* Strasbourg, 1993.

28. A. J. Landers, *Pesticide Outlook,* 1989, **1**, 27.

29. K-L. Nau, and H. Raffel, *Proc. Brighton Crop Protection Council Conference - Weeds,* 1991, 731.

30. A. R. Frost, *J.Agric.Engng Res.,* 1990, **46**, 55.

31. P. C. H. Miller and J. V. Stafford, *Proc. Brighton Crop Protection Conference - Weeds,* 1993, 1249.

32. M. E. R. Paice, P. C. H. Miller and J. D. Bodle, *J.Agric Engng Res.,* 1995, **60,** 107.

33. I. P. Craig, G. A. Matthews and E. W. Thornhill, *Crop Protection,* 1993, **12**, 549.

34. J. G. Elliott and B. J. Wilson, (Eds.) "The Influence of Weather on the Efficiency and Safety of Pesticide Application." *BCPC Occasional Publication 3,* 1983.

35. C. S. Parkin, A. J. Gilbert, E. S. E. Southcombe and C. J. Marshall, *Crop Protection,* 1994, **13**, 281.

36. D. A. Harris, K. S. Johnson and J. M. E. Ogilvy, *Proc. Brighton Crop Protection Council Conference - Weeds,* 1991, 715.

Pesticides in Foodstuffs

P. Smith

STRATHCLYDE REGIONAL CHEMIST'S DEPARTMENT, 64 EVERARD DRIVE,
GLASGOW G21 1XG, UK

1 INTRODUCTION

This paper is written from the Public Analyst's point of view. Public Analysts are essentially here to protect the public and, in their traditional role, particularly to protect the public from poor quality food, from the adulteration of food, and from misleading or fraudulent labelling of food. In Glasgow the service can be traced back to Mr Tatlock who was first appointed in 1875. In the past the profession was responsible for stopping such gross adulterations as the use of red lead to colour cheese, the resale of dried used tea leaves as fresh tea and the adulteration of coffee with acorns. As this type of gross adulteration has now been eliminated the public are, whether or not justified, becoming more sensitive to the presence of trace components in food. The infamous "e" additives, hormone growth promoters, antibiotics, and trace contaminants are now of concern. Reports published by organisations such as "Which" and "Mothers for Safe Food" illustrate this change.

Pesticides are, at the moment, essential in modern high production agriculture needed to feed the ever growing world population. However, traces of pesticides in the final product are generally considered to be unwanted contaminants. Legislation specifically to control pesticides is intended to reduce residues to a practical minimum mainly by the promotion of "Good Agricultural Practice" (GAP). Despite the precautions taken, residues do occur in and on food on retail sale in the UK. Some foods products and some pesticides are subject to quantitative controls through regulations. However, much of the food on sale in Britain is still not subject to any quantitative control of residue levels and the legislation that does exist is difficult to use at retail level.

2 LEGISLATION SPECIFIC TO PESTICIDES

The Food and Environment Protection Act of 1985, commonly referred to as FEPA, introduced the first direct controls of pesticides in Britain. The act is enabling legislation and Part 3 gives Ministers the power to :-

a. make regulations controlling the importation, sale supply and use of pesticides;

b. specify how much pesticide or pesticide residue may be left in crops food or feeding stuffs;

c. authorise any person to enforce the part of the act dealing with pesticides;

The FEPA powers were used to make The Control of Pesticides Regulations of 1986. These implement a and c above and prohibit the advertisement, sale, supply, storage and use of pesticides unless :-

a.   the pesticide in question has been approved for use;

b.   any conditions applied to the use of the pesticides are adhered to.

Essentially the conditions applied are intended to ensure the safe use of pesticides based on "good agricultural practice" (GAP). Pesticides must carry labels giving the conditions for safe use and GAP requires them to be used in accordance with the labelling conditions. These controls are intended to ensure, amongst other matters, that the residues remaining in the final foodstuffs are as low as is compatible with effective use.

The FEPA powers have also been used to make the Pesticides (Maximum Residue Levels in Crops, Food and Feeding Stuffs) Regulation of 1994. These implement the final point in the FEPA scope and lay down quantitative limits for the amounts of pesticide residues permissible in crops, in food for human consumption, and in animal feeding stuffs. These limits are set as "Maximum Residue Levels (MRLs)". MRLs are the maximum residues that should remain in products if GAP is followed. They are not directly related to toxicity or health but an MRL could not be set at a level which would not be safe for consumption. The regulations incorporate some EC MRLs set in Directives and also set other MRLs specific to the UK.

The regulations are very simple but, compared to the traditional food legislation, a little strange. The essential elements are :-

1.   Schedule 1 lists 85 primary compounds or groups of compounds to which the regulations apply.

2.   Schedule 2 Part 2 sets out a range of pesticide and product combinations and gives MRLs for a number, but not all, of these combinations. These combinations are taken from EC Directives and are intended to implement these Directives in the UK. The regulations also make it an offence to "put into circulation" any product failing to comply with the standards in this part of the schedule.

3.   Schedule 2 Part 1 sets out a different range of pesticide and product combinations and MRLs which are not subject to EC Directives. (i.e. these are UK specific MRLs). Any breach of this part would be an offence under section 16(12) of the Food and Environment Protection Act 1985.

These regulations give the only limits for pesticide residue levels in food in the UK.

3 MONITORING OF PESTICIDE RESIDUES

In the UK monitoring of foods for pesticides residues is carried out by Local Authorities as part of their duties under the Food Safety Act, and by the food industry for their own quality control. However most of this data is confidential and is not readily available. The most comprehensive data relating to the occurrence of pesticides in food is obtained by the Working Party on Pesticide Residues (WPPR). The data is obtained through four sub-groups "Fruit and Vegetables", "Cereals and Cereal Products", Products of Animal Origin", "Fish and Fish Products". Each undertake the surveillance of the appropriate categories of foods. The data obtained is published each year in the "Annual Report of the Working Party on Pesticide Residues" which also fulfils the UK obligation to report monitoring data to the EC. Consolidated reports are published every three years as Reports of the Steering Group on Chemical Aspects of Food Surveillance. Limited amounts of industry and local authority data are also included in these reports.

4 OVERALL MONITORING RESULTS

The published information gives a good picture of the levels of pesticides in UK food and the trends over the years. The staple foods of milk, bread and potatoes are examined every year supplemented by rolling programmes intended to cover most product and pesticide combinations over a more extended period. The overall results for the years 1989 to 1993 are given in Figure 1.

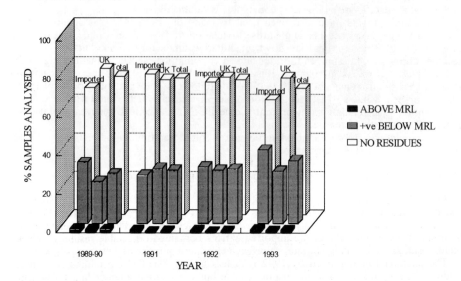

Figure 1 *Overall Monitoring Results 1989 to 1993*

The published information shows that between about 60% and 80% of samples examined do not contain detectable residues. Only 1 to 2% of samples contain any residues in excess of MRLs. When residues are detected they can usually be explained by :-

a. Pesticides used on the growing crop for the control of insects, moulds or other pests;

b. Pesticides used post harvest for the preservation of food during storage;

c. The carry-over of residues from the past use of persistent organo-chlorine compounds,

d. The current use, or abuse, of persistent organo-chlorine compounds (usually in imported products)

4.1 Pesticides Used on the Growing Crop

As an example of residues reaching food on retail sale due to the use of pesticides on the growing crop we can look at carrots.

A major pest of carrots is the carrot root fly. This is normally controlled by organophosphorus pesticides used on the growing crop. These commodities have not been examined every year but surveys of retail products were carried out in 1989-90 and in 1992. The incidences of residues in imported and UK produce are illustrated in Figure 2.

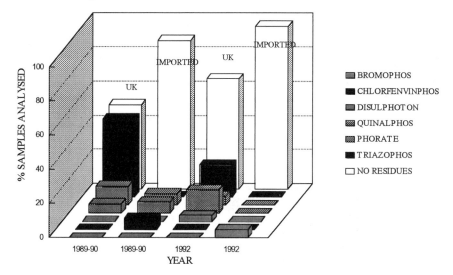

Figure 2 *Pesticides in UK and Imported Carrots 1989 to 1992*

In 1989/90 just over 50% of UK carrots contained a residue of one or more pesticides. Triazophos was the most common residue, occurring in almost 45% of samples. Phorate and quinalphos were the other major pesticides found presumably reflecting use patterns on the crops. Only about 19% of the imported carrots contained detectable residues. Where found, residues were mainly phorate, chlorfenvinphos and quinalphos.

The position had improved by 1992 with less than 40% of UK products containing residues. Triazophos was still the major residue but the relative amount had fallen with some increase in alternative pesticides. The position on imported carrots had also improved with over 95% of samples tested proving negative and just one sample containing bromophos.

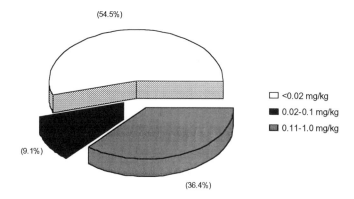

Figure 3 *Triazophos in UK Carrots 1989-90*

The distribution of the levels of residues of triazophos in the 1989-90 survey are given in Figure 3. The results are of some concern with 36% of the samples exceeding the MRL for triazophos, then set at 0.1 mg/kg by the Pesticides (Maximum Residue Levels in Food) Regulations 1988, by up to ten fold. The UK MRL was based on a draft Codex standard which was founded on incomplete data from trials in the 1970s. The Advisory Committee on Pesticides (ACP) suggested that a level of 2 mg/kg was more appropriate and called for more data. (If an MRL of 2 mg/kg had been in use, all samples would have complied with the standard).

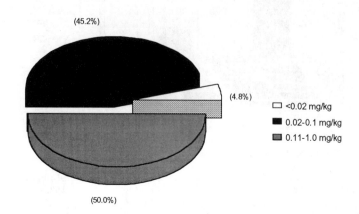

Figure 4 *Triazophos in Treated Carrots 1991*

As a follow up to these findings UK carrots of known treatment history were examined in 1991. These were commercially grown products which were treated in accordance with GAP for the pesticide(s) used. Figure 4 shows the findings in diagrammatic form. Over 95% of the carrots treated with triazophos were found to contain detectable residues with 50% of the samples containing residues in excess of the MRL of 0.1 mg/kg. (Smaller proportions of carrots treated with chlorfenvinphos and phorate also had residues in excess of the appropriate MRLs). Examination of treatment records indicated that approved procedures, i.e. GAP, had been followed and hence the MRL should not have been exceeded. These data indicated that the high incidence of triazophos found in UK carrots, and the high incidence of residues in excess of the MRL, were probably not due to misuse of the pesticide. The MRL in the regulations appeared to be incorrect.

The trade were informed of the findings and a second survey of retail carrots in 1992 showed improved results with less than 20% of UK carrots samples containing detectable triazophos. The proportion of samples containing residues in excess of the UK MRL had fallen dramatically though was still almost 12%. The changes were almost certainly due to changed practices following the release of the monitoring data to the trade. The drop in levels of triazophos was accompanied by an increased incidence of other pesticides though at levels below the MRLs.

A repeat of the surveillance of treated carrots in 1993 showed data similar to the retail studies of 1992. About 45% of carrots treated contained residues but those in excess of the MRL had fallen to about 12%.

Following this work a new UK MRL of 1 mg/kg was included in the 1994 regulations. All the results obtained over the period 1989-93 would have complied with this new standard. The information obtained during this work has shown the original UK MRL for triazophos in carrots to be quite significantly in error. It does lead to the question as to whether the trials data available from older pesticide is adequate to set appropriate MRLs and whether other examples of inappropriate standards lurk within the regulations.

4.2 Pesticides Used Post Harvest

Bread has been monitored regularly from 1984 but the form in which the data was presented was not standardised until 1988. A range of pesticides have been sought in this commodity but the only significant occurrences have been of a limited range of organophosphorus pesticides i.e. pirimiphos-methyl, chlorpyrifos-methyl, fenitrothion, malathion and etrimphos. The range of pesticides found correlates with the pesticides used post-harvest to prevent insect attack on stored grain. Figure 5 illustrates the distribution of residues found in all types of "brown" bread including wholemeal from 1984.

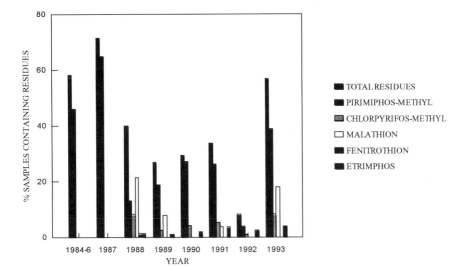

Figure 5 *Pesticide Residues in "Brown" Bread 1984-1993*

A similar range of pesticides has been found in white bread though the incidence of residues are generally more frequent in brown and wholemeal breads than in white. This is in line with monitoring of wheat products which indicates that residues are higher in bran and wheatgerm.

The only organochlorine pesticide detected has been gamma-HCH which was found in 10 out of 90 samples in 1988, 1 out of 148 samples in 1989, and 1 out of 188 samples in 1993.

The diagram shows a significant variation in the proportion of samples containing residues and in the specific pesticides detected from year to year. This may reflect varying risks of insect attack from year to year and/or varying sources of the wheat. In the period of 1990-92, summers were good in England resulting in more home grown wheat reaching bread making standards. This probably reduced the amount of imported wheat which had been treated in storage. It is known that malathion was used regularly in France while in the UK pirimiphos-methyl appears to be the pesticide of choice. A reduction in French wheat could account for the low incidence of malathion over 1990-92.

The data indicates quite a quite high incidence of residues in bread, particularly in brown bread. Data is available of the distribution of pesticide levels from 1990 onwards. By looking at this data for pirimiphos-methyl, the most common pesticide found, the significance of these residues is put into perspective. This data is shown in Figure 6.

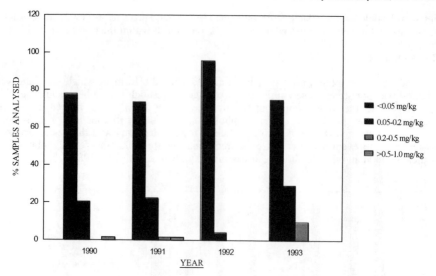

Figure 6 *Levels of Pirimiphos-methyl in "Brown" Bread 1990-93*

A large majority of samples of brown bread, around 70 to 80%, did not contain residues above the reporting limit of 0.05 mg/kg. The majority of the residues detected were at levels of 0.2 mg/kg or less and in only 2 years have any residues been detected above 0.5 mg/kg. In fact these were single samples each year and the actual results were 0.6 and 1.0 mg/kg.

Similar data has been produced for white bread but the incidence of residues was even lower exceeding 20% only in one year. In addition those residues that were found were at lower levels, with no samples found above 0.2 mg/kg.

These residues illustrate deficiencies in UK pesticides legislation. Residues are being regularly detected in bread, which is a staple diet item, but there are no UK MRLs applicable. The pesticides found are derived from use in the storage of wheat. They enter bread through the use of wheat containing pesticide residues in the manufacture of the flour which is the major ingredient of the food. There is a UK MRL of 5mg/kg applicable to pirimiphos-methyl in wheat and there are MRLs for other pesticides found regularly in wheat. In additive and contaminants regulations, made under the Food Safety Act, we would have carry over provisions which would allow us to calculate a standard for bread based on the amount of wheat it contained. According to the recipe I have for bread, this would set an acceptable residue level of about 3 mg/kg.

To complicate matters there are Codex MRLs of 1 mg/kg for wholemeal bread and 0.5 mg/kg for white bread. These are stricter than we would apply using carry over provisions which take no account of losses to be expected during baking.

The maximum residue regulations cannot be used to control pesticides in bread because the limits apply to primary foods only and there are no carry over provisions. It is at least encouraging that, whichever standard is applied, no sample of bread examined in the WPPR program since before 1988 has contained residues which would be considered excessive.

4.3 Extraneous Pesticides

Milk gives us a product which is not subject to pesticide treatment. Little data was available for milk prior to 1985 but data has been published by WPPR for each subsequent year. The data available before 1988 was published in a different format which is not directly comparable with later data. While a large range of pesticides of different classes has been sought, the only positive residues found have been the organochlorine pesticides alpha, beta and gamma-HCH, DDT and metabolites, hexachlorobenzene and dieldrin. Of these only dieldrin and gamma-HCH have been found at any significant frequency. Figure 7 summarises the occurrence of all residues, gamma-HCH and dieldrin.

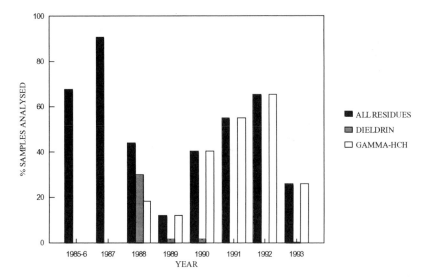

Figure 7 *Gamma-HCH and Dieldrin in Milk 1985-93*

The data available before 1988 does not allow the incidence of individual pesticides to be ascertained. However, in 1985-6 and 1987 the most frequent residue was dieldrin with gamma-HCH the next most common. This continued in 1988 after which the dieldrin incidence fell rapidly and positive findings are rare in later years. The most significant pesticide is gamma-HCH which has been found in every sample containing pesticide residues from 1989 to 1993. The incidence of residues reached a minimum in 1989 but appears to have been rising since. A big drop occurred in 1993 and it will be of interest to see if this is maintained when 1994 data becomes available later this year.

The apparent rise in the incidence of gamma-HCH residues appears, on first sight, to be disturbing. However, if we look at the actual levels of gamma-HCH, rather than the occurrence of detectable residues, a different picture emerges. This data is available only from 1990 and is presented in Figure 8. The highest incidence of occurrences are in the band around the reporting limit at up to about 10% of the MRL. The change in proportion of samples below the reporting limit and those just above match very well suggesting the apparent variations in incidence rate may well be due to minor fluctuations around this figure and have no real significance. The incidence of levels above about 10% of the MRL falls off very rapidly. In 1985-6 two samples exceeded the then EC MRL (there were no UK standards in existence at that time) but since then there has not been a single incidence of any pesticide in milk at or above the MRL found in surveillance work.

The source of residues in milk is open to some speculation. The residues do not appear to be due to any direct agricultural use but rather to adventitious contamination from past, or possibly current, uses. Some work has suggested the residues may be derived from animal feed as there is some indication that higher levels follow the use of feed supplements in winter. The very clear drop in the proportion of samples containing dieldrin is almost certainly due to the phasing out of this pesticide in the UK.

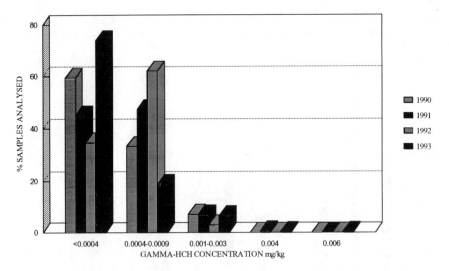

Figure 8 *Gamma-HCH in Milk 1990-93*

The limited scope of the pesticide regulations is again illustrated by the data obtained on milk. The MRL for gamma-HCH in milk now in force in the UK (0.008 mg/kg) would not strictly be applicable to the residues found because the wording of the regulations is such that MRLs apply only if the pesticide is :-

a. applied to the product or;

b. applied to the land on which the product was grown.

As the residues found do not arise from the direct use of the pesticide on milk or on the fields in which the cattle are kept, the MRL would not be enforceable.

4.4 Difficulties of Applying Maximum Residue Level Regulations

The examples discussed illustrate some of the deficiencies in the regulations controlling pesticides in retail products. The regulations are difficult to use for a number of reasons which include both those already indicated and other problems. These may be summarised as follows.

1. The products listed in the regulations are all, with the arguable exception of one group ("preparations of meat") primary foodstuffs. That is products such as cereals, fruit or meat as such, but not manufactured or value added foods. This means that they do not apply to the majority of food on sale. For example, they would apply to plums but not to plum jam.

2. There are no "carry over" provisions as found in most British additives and contaminants regulations.

3. Limits are not set for all the listed pesticides in all the listed products.

4. There are no general limits given for pesticides not specifically covered by the regulations (i.e. standards apply to 85 pesticides or groups only);

5. There are no general limits for the pesticides listed in foods other than those foods specifically given in the regulations.

6. There are no provisions to deal with multiple residues. In most additives regulations provisions are made for the levels permitted if multiple additives of similar classes are present.

7. There are sampling arrangements laid down in a rather loose way. Essentially they state that samples should be taken "so far as is practicable" using the procedure in the Codex document CAC/PR5-1984 Part 5 "Recommended Method of Sampling for the Determination of Pesticide Residues" . The procedures in this document are only applicable to consignments of produce and could not be used for products on retail sale in most cases.

8. The sampling procedures are incompatible with food sampling procedures laid down in the Food Safety (Sampling and Qualifications) Regulations 1990 (though they could be modified to comply if a sufficiently large sample was procured).

In a recent incident when DDT, well in excess of the MRL, was detected in English black-currants, no action was taken under the maximum residue level regulations. A case was taken under the COPR for illegal storage of a non-permitted pesticide. While this was a satisfactory outcome in some ways it was a rather circuitous route to take and would have failed if the pesticide had not been found on the premises.

4.5 Application of The Food Safety Act 1990

The composition and safety of food is controlled by The Food Safety Act 1990 and regulations made under this act. The general provisions of this act are very powerful and do give an alternative procedure for controlling pesticides in food. Little use has been made of the act to date as most of the provisions, while powerful, are difficult to use for pesticide residues. Some of the changes made in this act, when compared with the Food Act 1984 and earlier Food and Drugs Acts, do appear to make it more easily applied to the area of pesticide residues. The sections of concern to us in this case are Sections 7, 8, 14 and possibly 15.

4.5.1 Section 7, Food Injurious to Health. This section makes it an offence to render food injurious to health with the intent that it is to be sold for human consumption by any of the following processes :-

a. adding any substance to food,
b. use any article or substance as an ingredient in the preparation of food,
c. abstracting any constituent from food, or
d. subjecting food to any other process or treatment,

Where a pesticide has been added to food either on a farm or in post harvest treatment this subsection would be applicable. However to obtain a conviction two points must be proved :-

a. The operation has to render the food injurious to health. It would be necessary for a pesticide to be present at a very high level in order to prove unequivocally that the product was harmful to health even though the cumulative effect of an adulterant must be taken into account.

b. It is necessary for there to be an intention that the food shall be sold in that state. It would therefore be necessary to prove that the person responsible for treating the product intended it to be sold for human consumption in that state. It is difficult to see any circumstances, other than malicious contamination of the food or food source, when this point could be proved to the satisfaction of the court.

While this section is in theory very powerful, in practice it is little used.

4.5.2 Section 8, Selling Food not Complying with Food Safety Requirements. This section, which is a considerable change from the equivalent sections of earlier legislation, makes it an offence to sell, offer, expose, deposit, consign or advertise for sale for human consumption food which does not comply with food safety requirements. The section then states that food shall be taken as failing to comply if either :-

a. it has been rendered injurious to health by any of the operations described in section 7; or

b. it is unfit for human consumption; or

c. it is otherwise contaminated (whether by extraneous matter or otherwise) so that it would not be reasonable to expect it to be used for human consumption.

The comments regarding section 7 apply almost equally to 8(a), though using this subsection would apply at the point of retail sale and the act of selling is itself an offence without the necessity to prove intent. It would be easier to prove a case though it would still be necessary to prove that the food was harmful to health. Except in the case of direct sale from farm to the public the defence of warranty would be entered and further action would have to be taken against the producer with the problems already outlined. 8(b) is unlikely to apply to food treated with pesticides.

8(c) is very useful where a pesticide residue is found. It is only necessary to show that any residue found was such that it would "not be reasonable to expect it to be used for human consumption in that state". The problem is setting a standard which would apply to the residue. Where food is shown to contain a residue above the UK, EC or Codex MRLs a court may well accept there is an offence under this section. As far as I am aware this has not been tested in court to date though certificates have been written on this basis.

This section could also be used to control the presence of multiple residues, even where the individual residues were below appropriate MRLs, where effects were additive.

4.5.3 Section 14, Selling Food not of the Nature, Substance or Quality Demanded. This is the classic section of the Food Safety Act and dates back to the first practical legislation controlling food. It states that it is an offence to sell, to the purchaser's prejudice, food which is not :-

a. of the nature, or
b. of the substance, or
c. of the quality demanded by the purchaser.

Under this section there is no need to prove intent or harm to health. Where sale is to the prejudice of the purchaser an offence has been committed. It is only necessary to show that an inferior product was sold to the purchaser and not to prove that financial or other loss or damage was caused.

If a food was found on sale containing a pesticide and it could be shown that this rendered the food not of the nature, substance or quality demanded an offence would have been committed. I am aware of only one successful prosecution relating to pesticides and this involved "organically grown" food. A pesticide residue was found in food so described and the Public Analyst argued that the appropriate limit for pesticides in such food was nil and hence the food was not of the substance demanded. This interpretation was upheld by the court and a conviction was obtained. However, I understand that in a second similar case the court found for the defendant.

4.5.4 Section 15, Falsely Describing or Presenting Food. This section makes it an offence to give a label with food (or to advertise food) which falsely describes the food or is likely to mislead as to its nature, substance or quality. The regulation also controls the presentation of food.

The section would be applicable in a limited number of cases. For example, a food described as "free from pesticides" and found to contain residues.

When it comes to controlling pesticide residues in food at the retail level, the legal situation is not very satisfactory. The use of Section 8, and possibly section 14, of the Food Safety Act offers the best chance of control. However, to use the provisions of this act food analysts will have to convince the courts as to the standards to be applied. The sampling provisions of the maximum residue level regulations may complicate the use of MRLs as standards to be used in court.

5 SUMMARY

When looked at from the point of view of a Public Analyst or an authorised officer of a food authority, the legislation applicable directly to food is weak. The question we should ask is "does this matter"? Does the current legislation effectively control residues in food by controlling the use of pesticides? The data available suggests about one third of basic and primary foodstuffs contain detectable residues with only 1 to 2% containing levels in excess of statutory limits. The numbers of residues sought have increased over time which implies the incidence of residues is slowly falling. These limits are set on the basis of good agricultural practice and an excess does not necessarily imply a health hazard. My personal opinion is that legislation should be enforceable and good legislation would ensure the protection of both the public and the reputable producer and trader.

6 ACKNOWLEDGEMENTS

Data in this paper has been taken from the following reports :-

a. Food Surveillance Paper No. 25, Report of the Working Party on Pesticide Residues 1985-88.
b. Food Surveillance Paper No. 34, Report of the Working Party on Pesticide Residues 1988-90.
c. Annual Report of the Working Party on Pesticide Residues 1989-90.
d. Annual Report of the Working Party on Pesticide Residues 1991.
e. Annual Report of the Working Party on Pesticide Residues 1992.
f. Annual Report of the Working Party on Pesticide Residues 1993.

All are published by HMSO.

The Poisoning of Non-target Animals

K. Hunter

SCOTTISH AGRICULTURAL SCIENCE AGENCY, EAST CRAIGS, CRAIGS ROAD,
EDINBURGH EH12 8NJ, UK

1 INTRODUCTION

Pesticides are used extensively in modern arable farming and to a lesser extent in other sectors of food production. Some pesticides are also used in areas unrelated to agricultural production, eg non-agricultural use of herbicides in weed control, rodent control in urban areas, and the use of insecticides to counter public health concerns. The usage of pesticides on field crops in the United Kingdom (UK) has been estimated to be in excess of 30,000 tonnes of active ingredients per annum[1.] Scrutiny of basic toxicological data for active ingredients suggests that some pesticides could exhibit potential hazards to non-target animals however such hazards will not necessarily be realised when formulated products are used in the field.

In the UK, schemes to investigate incidents of animal poisoning that result from the use of pesticides are operated by the Agricultural Departments of central government. These so called 'Wildlife Incident Investigation Schemes' have their origins back in the early 1960s at a time when there were great concerns about the poisoning of wildlife by the extensive use of organochlorine insecticides.

The schemes have evolved since those days to fulfil a surveillance role and to play a part in the regulatory process which controls the supply and use of pesticides. With the introduction of the Control of Pesticides Regulations in 1986 the Agricultural Departments took on direct enforcement responsibilities in respect of incidents involving either deliberate abuse of pesticides or misuse of pesticides.

During the development of new products commercial companies invest a great deal of effort to ensure that candidate products are not only effective for their intended use but are also acceptable in terms of their environmental impact on non-target species. As part of this any evidence of adverse effects on non-target animals is assessed in carefully supervised field trials. Following application, the registration body carries out a risk assessment process on the proposed uses before deciding whether or not it would be environmentally safe to grant some level of approval. Despite this scrutiny it is possible that some animals may suffer adverse effects when a product is used on a wider basis, under a varied range of conditions. It is obviously impossible to test products for hazards to all species and past experience has shown that biochemical and metabolic differences between species can give rise to selective toxic effects. A good example of this was the poisoning of *Anser* geese by carbophenothion[2].

It is in this area that the WIIS schemes carry out their surveillance activities and in doing so complete a feedback loop to the regulatory function. Data gathered in this way can be used to support a) the confirmation or extension of approval status when no significant adverse effects have been revealed; b) modification of the conditions of use where this is likely to minimise a hazard that has been revealed; and c) in an extreme (and unlikely) case the revocation of approval status. The data can also provide useful comparisons in risk assessment on other products or can serve to trigger further detailed studies in specific areas.

If data on animal poisoning is to be used in this way it is crucial to gather relevant field information about the circumstances of exposure. Equally at post-mortem examination it is important to obtain any pathological evidence that might support a diagnosis of exposure. These stages can also act as filters in rejecting incidents that do not justify analytical investigation. Residues of pesticides in poisoned animals may be relatively low hence analytical methodology employed in the analytical investigation must be sensitive and robust. As in all trace analysis it is vital that rigorous confirmation criteria are applied to any residues identified.

On limited occasions there may be valid reasons to override the rejection mechanisms as opportunities to carry out specific surveillance arise. Such surveillance would normally be restricted to certain combinations of species and class of pesticide where there are particular concerns.

The schemes are reactive in nature and their success depends on members of the public and interested organisations to notify incidents. Accordingly, as well as satisfying reporting requirements associated with the regulatory and enforcement aspects it is appropriate to ensure some positive feedback on the outcome of investigations to the original notifiers of each incident.

2 RESULTS OF INCIDENT INVESTIGATIONS

The scope of the schemes is not restricted to wild animals as their titles suggest. At present they cover wild birds, wild mammals, companion animals such as cats and dogs, farm livestock and beneficial insects (essentially honeybees). Up to 800 potential incidents may be notified each year throughout the UK[3]. An incident may involve a single animal or a number of animals of the same or different species.

In Scotland an average of 210 incidents was notified per annum for the years 1991-94, the animal categories involved in these incidents are shown in Figure 1. Mortalities involving birds were most frequently notified (47.5%). This may be due in part to a great public interest in birds but also reflects a relatively greater chance of discovery than for many mammals. However assessment of the reporting frequencies of ringed birds would suggest that the sample of birds received will have been biased towards the larger species which are more readily found. Mortalities of companion animals feature prominently (28.5%), no doubt in part at least because their welfare is closely associated with specific owners. At the same time many of these animals enjoy periods when little or no direct control is exercised over their activities.

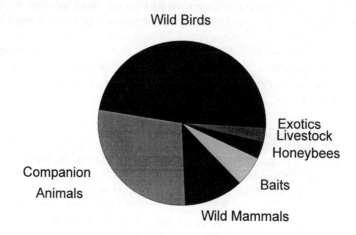

Figure 1. *Incidents Investigated in Scotland 1991-1994, Animal Categories.*

Examination of the outcome of investigations into these incidents shows that the cause of death was confirmed as being due to pesticide poisoning in only 22% of notified incidents (Figure 2). Other causes of death such as starvation, disease and trauma associated with road traffic accidents or gunshot wounds accounted for a similar proportion of incidents. A very small number of non-pesticide poisonings were included in this category. A large proportion of investigations were terminated with no definite cause of death having been established, since further examination was unlikely to be cost effective.

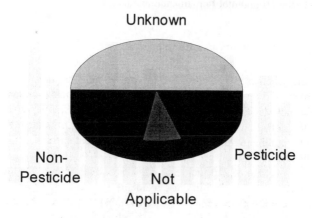

Figure 2. *Cause of Death in Wildlife Incidents in Scotland 1991-1994*

A preliminary assessment of the underlying causes of pesticide poisoning can be obtained by classifying incidents of confirmed poisoning into simple categories by use (Figure 3). The major problem is then clearly identified with the deliberate abuse of pesticide products to poison animals. In the last 4 years in Scotland 73% of incidents reflected this type of illegal practice. Throughout the UK as a whole incidents involving deliberate abuse have made up 58 to 70% of all poisoning incidents in recent years.

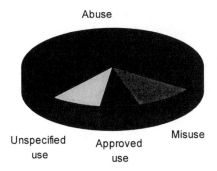

Abuse

Unspecified   Misuse
use   Approved
use

Figure 3. *Pesticide Incidents in Scotland 1991-1994*

Misuse of products, by either a lack of care or a failure to adhere to correct practice caused 11% of confirmed incidents. Reassuringly, only 7% of incidents were attributed to the approved use of pesticide products. In a small proportion of incidents it proves impossible to reliably ascribe them to a single specific category and these are simply denoted as unspecified use.

The perennial nature of the abuse problem is demonstrated in Figure 4. The difference in number of incidents from year to year is more likely to be a reflection of variability in reporting frequencies than any other specific factor. This situation prevails despite the introduction of a co-ordinated Anti-Abuse Campaign in 1991 by Government departments and other organisations.

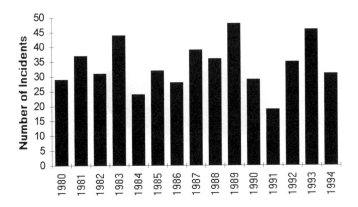

Figure 4. *Pesticide Abuse in Scotland*

The most common victims of abuse in Scotland are birds of prey such as buzzards or golden eagles but the list extends to cover most species including rare visiting birds and those subject to re-introduction programmes such as the red kite and the white tailed eagle. Companion animals are other frequent casualties of deliberate abuse, both in rural areas and also in urban areas.

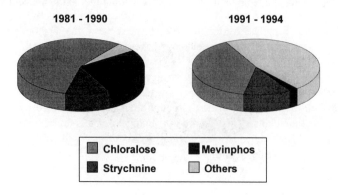

Figure 5. *Pesticide Abuse in Scotland*

In the 1980s, and the years preceding that decade, abuse in Scotland was almost exclusively restricted to 3 chemicals (Figure 5). These were chloralose, a narcotic rodenticide approved for the control of mice, indoors only; strychnine which is used exclusively for mole control; and mevinphos, one of the more toxic of the organophosphorus insecticides. In the first part of the current decade there has been a major reduction in the extent of abuse of mevinphos. This coincides with the effective removal from supply by the manufacturers and the subsequent termination of the approval status. The most profound change however was a dramatic rise in attempts to abuse a range of other pesticide formulations.

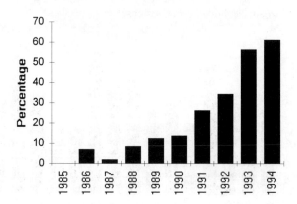

Figure 6. *Proportion of Vertebrate Incidents in Scotland Associated with the Abuse of Pesticides, other than Chloralose, Mevinphos and Strychnine.*

Although the numbers of incidents of abuse of specific chemicals varied from year to year there was a steady rise in the proportion of incidents associated with the abuse of chemicals other than chloralose, mevinphos and strychnine, to the extent that such incidents made up 63% of vertebrate abuse incidents in 1994 (Figure 6). A range of products including, insecticides, rodenticides and molluscicides have featured but there is no doubt that this fundamental change in pattern is closely linked with the abuse of toxic carbamate insecticides and especially with the extensive abuse of formulations containing carbofuran. The Agricultural Departments treat the abuse of pesticides very seriously and seek to bring offenders to court where ever possible.

The sheer magnitude of the abuse problem tends to mask any general patterns linked more closely with the normal use of pesticides.   A different insight into the problems associated with the poisoning of non-target animals can be obtained by removing data on abuse related incidents. This is valid in the context of this presentation because most of the incidents attributed to misuse arise in circumstances that are close to normal agronomic practice in the field although clearly they are not consistent with all of the conditions of use. Similarly it would seem likely, from the limited evidence available, that many incidents which fell in the unspecified use category originated from exposures that related to approved use or misuse of the relevant product. The only qualification is that a small proportion of incidents categorised as resulting from unspecified use may have been caused by deliberate abuse.

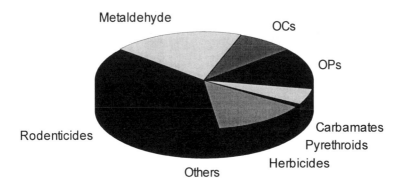

Figure 7. *Vertebrate Pesticide Incidents in the UK 1990-1994 (Data exclude abuse)*

In Figure 7, vertebrate poisoning incidents from the whole of the UK have been selected in this way and sorted on the basis of the type of pesticide involved. In the 5 year period 1990-94 there was a total of 318 confirmed incidents. A third of these were caused by exposure to anticoagulant rodenticides, 20% by metaldehyde and a further 30% by the various classes of insecticide. The remainder were mostly associated with the use of herbicides and virtually all involved some form of exposure to formulations containing paraquat.

It is perhaps not surprising that anticoagulant rodenticides feature so prominently, since they are designed specifically to poison small mammals and the most common form of bait material is cereal-based and potentially attractive to many non-target animals. Residues of the second generation rodenticides are relatively persistent in the tissues of exposed animals and this can lead to a risk of secondary poisoning. Rodents or other small mammals that gain access to baits may be more vulnerable to attack by predators during the period between ingestion and eventual death, which can take up to several days. The carcases of poisoned animals would in turn become sources of food for carrion feeders.

The distributions of rodenticide incidents in Scotland and in England & Wales were broadly similar (Figure 8), the minor differences largely reflected regional variations in the pattern of usage. Relatively fewer incidents were caused by brodifacoum, the most toxic of the rodenticides. Rodenticide formulations based on brodifacoum are subject to greater restrictions of use. Many incidents resulted from exposure to two of the more toxic rodenticides, bromadiolone and difenacoum, however a significant proportion resulted from exposure to the older and less toxic rodenticide, coumatetralyl.

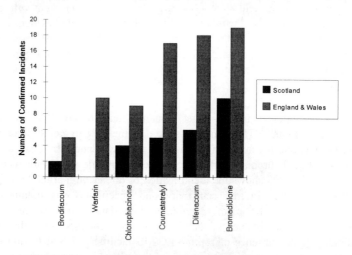

Figure 8. *Anticoagulant Rodenticide Incidents 1990-1994 (Data exclude abuse)*

In Table 1 the relative distribution of rodenticide incidents in Scotland is contrasted with estimates of the usage of this type of product in Scottish agriculture[4,5]. There is some correlation between the proportion of incidents involving bromadiolone and chlorophacinone (and to a lesser extent for difenacoum) and the extent of use of bait material containing these chemicals. There is a similar correlation between the use of active ingredients and incidents involving difenacoum and chlorophacinone. The anomalous results for coumatetralyl can be explained by the fact that this substance is used principally in tracking dust formulations. Such formulations contain a much higher concentration of active ingredient than conventional bait materials.

The main casualties of rodenticide poisoning were companion animals, seed eating birds, birds of prey and wild mammals such as foxes and badgers. In some cases the exposure was associated with direct ingestion of bait material; in other cases secondary poisoning was the most likely route. Animals at risk from secondary poisoning are unlikely to take in a toxic dose from a single prey item; however residues can accumulate in the liver, where the metabolic clearance rate is low, until a toxic threshold is exceeded and symptoms of poisoning begin to be expressed.

Table 1, *Anticoagulant Rodenticide poisoning incidents (1991-94) and rodenticide usage in Scotland.*

Rodenticide	Incidents	Usage of Bait	Usage of active ingredient
Brodifacoum	7.4%	0.2%	0.05%
Warfarin	0	2.4%	6.2%
Chlorophacinone	14.8%	14.8%	12.6%
Coumatetralyl	18.5%	2.8%	53.3%
Difenacoum	22.2%	40.1%	20.5%
Bromadiolone	37.0%	39.7%	7.5%
TOTALS	27	163.7 Tonnes *	26.6 kg *

*Estimated annual usage

   Residues of rodenticides in non-target casualties tend to be very low and are concentrated in liver tissue[6]. The compounds are not amenable to gas chromatographic analysis and the laboratories involved in this type of surveillance have had to exploit high-performance liquid chromatographic techniques to provide a basis for their multi-residue analysis[7]. Figure 9 shows the identification of difenacoum at 0.09 mgkg^{-1} in a barn owl liver. The required sensitivity and specificity was obtained using fluorescence detection. It is of paramount importance to confirm that the chromatographic response obtained in such instances truly corresponds to the presence of a rodenticide. In the absence of routine LC/MS facilities, UV detection used in series may offer some assistance. Examination of the relevant chromatogram (Figure 9) in this particular case showed a peak at the correct retention time and in the appropriate ratio compared to an authentic analytical standard of difenacoum. The chromatogram also revealed that, despite efforts to clean-up the extract, many other components were present and confirmation on this basis alone was not entirely convincing. However using diode array technology it was possible to examine the full UV spectrum of this component and compare it with that of the reference material. When this was done it was possible to get a good quality spectral match even at this very low residue level.

   This particular analytical application not only serves to indicate the technical demands placed on the analyst in this type of work but it also illustrates 2 further points. Firstly this incident is an example of the value of specific surveillance, the bird in question appeared to be a victim of a road traffic accident. The screening was only carried out because of present concerns about the effects of rodenticides on owls [8, 9]. Secondly it also serves to illustrate a difficulty in interpretation of residue data. The magnitude of the residue falls into a range where it could represent sub-lethal exposure

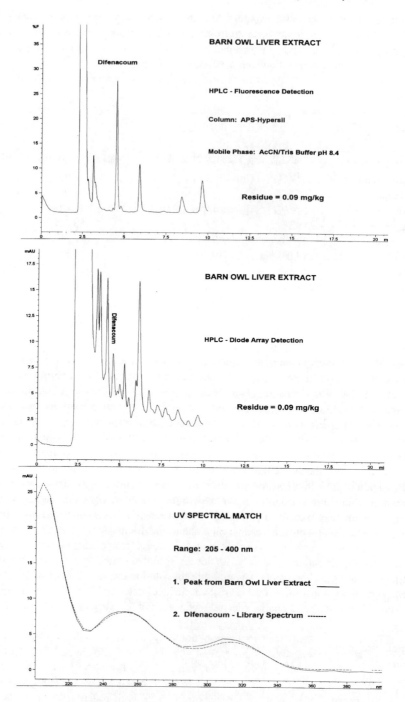

Figure 9. *HPLC determination of Difenacoum using fluorescence detection (top), diode array detection (middle), and spectral confirmation (bottom).*

for owls; on the other hand haemorrhagic symptoms have been expressed in other animal casualties where the residues of rodenticides in liver tissues were lower.

Table 2, *Vertebrate Incidents Involving Carbamate Insecticides, UK Data 1990-1994 (excluding abuse)*

Sources		Animals Affected
Granular applications	(2)	Blackbird, sparrow, pheasants, gull
Secondary poisoning	(1)	Buzzard
Seed treatments	(3)	Corvids, duck, dog
Spillage of molluscicide	(2)	Dogs
Unknown	(7)	Buzzard, peregrine, corvids, starlings, fox, cat, dog

The specific active ingredient associated with the most incidents (62) was metaldehyde. This substance is formulated in bran-based pellets as a molluscicide to control slugs. The bait material is very attractive to many animals and this was probably a major factor in the relatively high proportion of poisonings of non-target animals by this material. Only 11% of confirmed metaldehyde poisonings were attributed to approved use of products. In contrast over 50% of the incidents resulted from misuse, usually in circumstances where significant spillages occurred at the site of application or where animals gained access to bulk materials because of poor storage conditions.

The carbamate insecticide methiocarb is also formulated in bran-based formulations to control slugs. Pesticide usage data indicate that the area treated with methiocarb for this purpose in arable farming is as extensive as that for metaldehyde[1]. Methiocarb itself is more toxic than metaldehyde and formulations of methiocarb contain a higher proportion of a toxic dose (LD_{50}) per pellet than metaldehyde formulations. It may then be surprising that very few incidents resulted from the use of methiocarb as a molluscicide (Table 2). However specific studies have shown that methiocarb used in this way is capable of causing dramatic, short-term, local reductions in the populations of small mammals such as field mice[10]. Re-invasion by mice from surrounding untreated areas quickly restore populations back to the original level.

Table 3, *Vertebrate Incidents Involving Organophosphorus Insecticides, UK Data 1990-1994 (excluding abuse)*

Sources		Animals Affected
Agricultural sprays	(4)	Ducks, geese, dog, cattle
Improper disposal	(3)	Gulls, dog, cattle
Poor storage	(1)	Dog
Seed treatments	(19)	Pigeons, pheasant, corvids, geese
Veterinary medicines	(11)	Buzzard, corvids, parrot, poultry, dogs
Unknown	(8)	Corvids, geese, pigeons, pheasant, starlings, dogs, poultry

Poisoning of non-target animals by organophosphorus insecticides made up almost 15% of confirmed vertebrate incidents (Table 3). Scrutiny of the sources of the materials involved reveals that 2 areas featured most regularly. The first of these was a large group of incidents associated with seed treatments on cereal seed, the particular active ingredients being fonofos and chlorfenvinphos. The balance of evidence suggested that some of these incidents arose from approved use but in other cases the incidents were associated with significant spillages of seed in the field and hence these constituted misuse. A second large group of incidents resulted from the use of a number of organophosphorus insecticides as veterinary medicines.

Table 4, *Vertebrate Incidents Involving Organochlorine Insecticides, UK Data 1990-1994 (excluding abuse)*

Sources		Animals Affected
Amenity use-lumbricide	(4)	Barn owl, corvids
Improper disposal	(1)	Otter
Seed treatments	(2)	Pigeons
Timber preservatives	(2)	Bats
Veterinary medicines	(5)	Cockateel, cats, dogs
Unknown/Bioaccumulation	(12)	Kestrel, merlin, owl, sparrowhawk, cat

Although the use of organochlorine insecticides has been severely reduced over the years they were still the cause of a small proportion (8%) of poisoning incidents. A definite source of exposure could not be identified in many instances (Table 4) and these incidents probably reflect bioaccumulation from a variety of sources. The persistence of this class of pesticide results in occasional incidents associated with former uses, as timber preservatives for instance. A limited number of incidents resulted from the use of chlordane formulations to control earthworms on amenity areas such as golf courses.

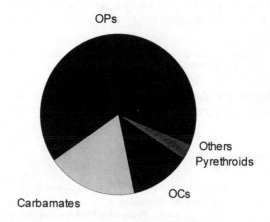

Figure 10. *Honeybee Poisoning Incidents in the UK 1990-94*

Monitoring honeybee incidents is important not only because of the role of bees as pollinators and their economic value, but because they provide a useful indicator species for hazards to beneficial insects in general. The fact that they are managed by man means that effects on their well being are noticed whereas the impact of normal agrochemical applications on other insects is likely to go unnoticed. In the last 5 years a total of 129 incidents of honeybee mortality resulted from exposures to pesticides. As might be expected, this type of incident was almost exclusively associated with insecticide formulations (Figure 10). The majority of incidents (65%) resulted from exposure to organophosphorus insecticides, with smaller proportions being caused by exposure to carbamate or organochlorine insecticides. In general terms many incidents were associated with applications to flowering crops such as oilseed rape, field beans and raspberries. Factors such as the timing of application and a lack of liaison with beekeepers were important facets in many incidents. Small numbers of incidents resulted from attempts to control feral bees with carbamate insecticides, from the use of organochlorine insecticides as timber preservatives, and from the malicious use of insecticides in isolated instances of deliberate abuse. These data are broadly similar to those reported in a detailed review of honeybee poisoning in Great Britain for 1981-1991[11].

3 SUMMARY

Incidents of poisoning of non-target animals by pesticides are investigated by government schemes in the UK. The reactive mechanisms employed are effective for dealing with acute toxic effects which are readily apparent in the field, but are less good at identifying chronic, and sometimes sub-lethal, effects. It has to be recognised that there will be some degree of under reporting of incidents coupled with some bias in the sample of incidents actually reported.

The major problem of poisoning of non-target animals in recent years has continued to result from the deliberate abuse of pesticides. Although the pattern of pesticides subject to abuse has tended to change there has been no evidence to suggest any change in the actual extent of abuse.

If abuse is excluded from the poisoning of non-target vertebrates the incidents can be linked with specific classes of pesticides such as anticoagulant rodenticides, metaldehyde and insecticides. Few of these compounds appear in lists of the 50 most extensively used active ingredients by either weight or area treated.

Acknowledgement

The data used in this presentation were produced by the Wildlife Incident Investigation Schemes operated in England & Wales by CSL and ADAS on behalf of MAFF, in Northern Ireland by DANI, and in Scotland by SASA on behalf of SOAFD.

Reports on the poisoning of animals by pesticides in the UK are published annually on behalf of the Environmental Panel of the Advisory Committee on Pesticides. They are available from the Agricultural Departments or from MAFF publications, London.

References

1. R. P. Davis, M. R. Thomas, D. G. Garthwaite and H. M. Bowen (1992). Pesticide Usage Survey Report 108 - Arable Crops 1992, MAFF London.

2. G. A. Hamilton, K. Hunter, A. S. Ritchie, A. D. Ruthven, P. M. Brown and P. I. Stanley, *Pestic. Sci.*, 1976 **7**, 175.

3. M. R. Fletcher, K. Hunter, and E. A. Barnett (1994). Pesticide Poisoning of Animals 1993: Investigations of Suspected Incidents in the United Kingdom, MAFF London.

4. L. Thomas (1993). Rodenticide Usage on Farms Growing Arable Crops in Scotland 1992, Scottish Agricultural Science Agency, SOAFD, Edinburgh.

5. L. Thomas (1994). Rodenticide Usage on Farms Growing Fodder & Forage Crops in Scotland 1993, Scottish Agricultural Science Agency, SOAFD, Edinburgh.

6. K. Hunter, *J. Chromatogr.*, 1983, **270**, 267.

7. K. Hunter, *J. Chromatogr.*, 1985, **321**, 255.

8. I. Newton, I. Wyllie, and P. Freestone, *Environ. Pollut.*, 1990, **68**, 101.

9. C. V. Eadsforth, A. J. Dutton, E. G. Harrison and J. A. Vaughan, *Pestic. Sci.*, 1991, **32**, 105.

10. I.P. Johnson, J.R. Flowerdew, and R. Hare, *Bull. Environ. Contam. Toxicol.*, 1991, **46**, 84.

11. P. W. Greig-Smith, H. M. Thompson, A. R. Hardy, M. H. Bew, E. Findlay and J. H. Stevenson, *Crop Protection,* 1994, **13**, 567.

Assessing the Environmental Impact of Sheep Dip on Surface Waters in Northern Ireland

R. H. Foy,[1] C. Clenaghan,[2] S. Jess,[3] and S. H. Mitchell[4]

[1]AGRICULTURAL & ENVIRONMENTAL SCIENCE DIVISION, DEPARTMENT OF AGRICULTURE FOR NORTHERN IRELAND.
[2]DEPARTMENT OF AGRICULTURAL & ENVIRONMENTAL SCIENCE, QUEENS UNIVERSITY BELFAST.
[3]APPLIED PLANT SCIENCE DIVISION, DEPARTMENT OF AGRICULTURE FOR NORTHERN IRELAND.
[4]FOOD SCIENCE DIVISION, DEPARTMENT OF AGRICULTURE FOR NORTHERN IRELAND.

NEWFORGE LANE, BELFAST BT9 5PX, UK

1 INTRODUCTION

There are two main periods for dipping sheep in the British Isles: a summer dip for controlling larvae of the blow fly (*Lucilla* and *Calliphora* spp.) and an autumn dip to control the causative agent of sheep scab (*Psoroptes communis* Hering, the sheep mite). Other parasites such as lice (*Damanalia ovis* Schrank and *Linognathus ovillus* Neum) and keds (*Melophagus ovinus* Linnaeus) can also be controlled by these dips while a spring dip may be required for the control of sheep tick (*Ixodes ricinus* Linnaeus). Active ingredients in dips typically consist of phenols plus one or more additional insecticides. Organochlorine (OC) compounds in dips were phased out from 1965 to 1984 to be replaced by organophosphorus (OP) compounds which are less persistent in the environment. A synthetic pyrethroid, flumethrin, is now marketed as an OP alternative, primarily as a response to concerns as to the effects of OP compounds on the operators of sheep-dips.[1]

The contamination of surface and groundwaters by the incorrect disposal of waste or surplus dip has been of long standing concern. Hynes noted the adverse impact of sheep dip containing BHC on the invertebrate fauna of a small stream in the Isle of Man.[2] A 1965-66 survey of sheep farms in Britain estimated that 13% of dips discharged to a ditch or permanent water.[3] Scottish surveys in 1973 and 1978 found that between 5% and 6% of dips discharged to a ditch or stream with 59% and 40% respectively discharging to soakaways, creating a potential hazard to ground water.[4,5] Similar percentages where found in a 1981 survey of dips within Northern Ireland where 10.4 % of dips discharged to a stream or ditch and 36.3 % to a soakaway.[6]

Recent studies have recorded a widespread occurrence of OPs in some Scottish streams and rivers during the autumn dipping season despite low numbers of recorded incidents of water pollution attributable to sheep dips.[7,8] The present paper describes the results of investigations on the impact of dips on water-courses in Northern Ireland which were prompted by increasing sheep numbers and evidence of poor disposal practices. Results are presented for dip usage and disposal, river monitoring for pesticides and biological surveys of stream invertebrates. As in other regions of the United Kingdom, an obligation to dip sheep, particularly for the control of sheep scab, was enforced in Northern Ireland by legislative measures. In June 1993 the legislation was amended so that compulsory dipping was no longer required.

88

2 STUDY AREA

The sheep population in Northern Ireland expanded from 1.06×10^6 in 1980 to 2.51×10^6 in 1994 and in upland areas they constitute the main livestock type.[9] The headwaters of the Upper Bann catchment, which drains part of the Mourne Mountains was selected for a survey of stream macro-invertebrates and water chemistry (Figure 1). This study area was also included in a UK investigation of the impact of sheep dip disposal in 1991 when stream and soil monitoring of pesticides were undertaken.[10] The underlying geology of the area is a mixture of granitic and Silurian rocks and the area has been subject to a study of the impact of agriculture on water quality.[11,12] The Upper Bann is in turn one of six major rivers draining into Lough Neagh which were sampled for pesticides in 1988-90. In total, these rivers account for 30% of the land area of Northern Ireland, and the agriculture in the catchments reflects that of Northern Ireland.[13]

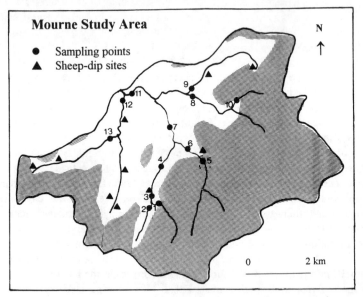

Figure 1 *Mourne study area showing sampling points and dip sites. Shaded area denotes land over 200m*

3 METHODS

3.1 Survey of sheep-dip pesticides and disposal methods

A survey requesting details of dip usage and disposal from 471 sheep farms was carried out in 1988 using the sampling methodology of the 1981 survey.[6] Farms were identified from the Northern Ireland Agricultural Census and the sample stratified according to flock size.

3.2 Pesticide analyses

Determination of OC and OP compounds present in the six major rivers entering Lough Neagh were undertaken using previously described procedures.[14,15] Snap samples were taken from the rivers, Six Mile Water, Main, Moyola, Ballinderry, Blackwater and Upper Bann, at locations close to where they entered Lough Neagh.[13] Ten samples were taken from each river at approximately 3 monthly intervals between 1988 and 1990.

3.3 Stream macroinvertebrate and water sampling

From October 1991 to December 1994 tri-annual sampling of stream macro-invertebrates and water chemistry was carried out at the 13 sites shown in Figure 1. Sheep dips were not in operation upstream of sites 1,2,3,5,8 and 10 during the study. Sampling took place during April/May, August (after summer dip) and late October-December (after autumn dip). Dissolved oxygen saturation (%DO) and temperature were measured on site. Water samples collected in acid-washed Nalgene screw top 2l bottles were analysed for nitrogen and phosphorus fractions, major ions, pH and conductivity on return to the laboratory.[11]

Stream macroinvertebrates were collected by kick sampling to a substrate depth of at least 5 cm, sweep sampling and stone washing into a standard pond net (mesh size 0.9 mm) for three minutes. Samples were preserved in 5% formaldehyde solution, hand-sorted in the laboratory and invertebrates present identified to family level. Biotic scores were determined according to the scheme proposed by the Biological Monitoring Working Party (BMWP) which assigns to most families of stream invertebrates a value on a 1 - 10 scale depending on their sensitivity to organic pollution.[16] The sum of BMWP scores was computed to give the BMWP biotic score.

3.4 Statistical analyses

The significance of the difference between the mean values of two data sets was assessed by a t test, preceded by an F test for the equality of population variances.[17] In order to assess spatial variation between the Mourne sites, stream macroinvertebrate and chemical data were each subjected to cluster analysis using Bray-Curtis similarity indices[18] with associated dendrograms.[19] The data were divided into two periods: before and after June 1993, when the compulsory dipping of sheep ceased. Biological data consisted of site specific presence and absence data for each stream invertebrate family by season so that, for each site, the number of seasons when each family was encountered was calculated.

4 RESULTS

4.1 Sheep dip usage and disposal.

Despite an increase in sheep numbers of 103 % between 1980 and 1988,[9] the sheep-dip usage survey showed a reduction in active ingredients between 1981 and 1988 (Table 1). In percentage terms the reduction was greatest for phenols which declined by 43% compared to 19% for the combined usage of OP and OC compounds. By 1988 OCs were no longer in use and the number of OPs had declined from nine in 1981 to three. Diazinon and propetamphos accounted for 94% of OP usage in 1988 compared to 4% in 1981. The amounts of dip used at the different dip periods remained relatively unchanged between 1981 and 1988, with the spring dip accounting for only a small percentage of total consumption (Table 2).

Table 1   *Active ingredients in sheep dip 1981 and 1988. Percentages refer to
sum of organochlorine and organophosphorus compounds*

		1981		1988	
Compound		tonnes yr^{-1}	%	tonnes yr^{-1}	%
Phenols		21.64		12.3	
Organochlorine	HCH	2.95	31.9	0	0
Organophosphorus	Diazinon	0.59	6.4	3.49	46.5
	Propetamphos	0.00	0.0	3.56	37.4
	Chlorfenvinphos	3.69	39.7	0.45	6.0
	Carbophenthion	0.97	10.4	0	0
	Dichlorfenthion	0.55	6.9	0	0
	Dioxathion	0.35	3.8	0	0
	Chloropyrifos	0.17	1.8	0	0
	Fenchlorphos	0.01	>0.1	0	0
	Iodophos	>0.01	>0.1	0	0
	TOTAL	30.93		19.8	

Table 2   *Seasonality of use of active ingredients in sheep dip 1981 and 1988.*

Dip period	1981 %	1988 %
Spring	4.8	5.5
Summer	54.5	50.1
Autumn	40.7	44.4

There was a marked reduction in dip operators who admitted to discharging to a ditch or
stream; from 10.4 % in 1981 to 0.6% in 1988 (Table 3). However, the category of disposal
to surrounding soil or permanent pasture included in the 1988 survey could include sites
where the waste was allowed to run freely away from the dip rather than being actively
spread. Soakaways remained in use in over one third of sites in 1988.

Table 3   *Method of disposal of waste or surplus dip (nd = not defined in 1981 survey)*

Category	1981 %	1988 %
Soakaway	36.3	34.1
Surrounding soil	40.1	17.1
Permanent pasture	nd	34.7
Ditch/stream	10.4	0.6
Slurry tank	nd	7.7
Other	13.1	4.9

4.2 River Pesticide Concentrations

A summary of pesticide monitoring in the Lough Neagh major rivers is given in Table 4. Of the OP compounds employed in dips, chlorfenvinphos was not actively monitored while propetamphos did not occur above the 5 ng l⁻¹ detection limit. Diazinon was observed above the detection limit on one occasion at a concentration of 110 ng l⁻¹ which exceeded the 100 ng l⁻¹ maximum allowable concentration (MAC) for individual pesticides as defined in the EC Drinking Water Directive. This exceedance occurred towards the end of the compulsory autumn dip period (14 November 1989) in the Six Mile Water. Diazinon was one of only two chemicals which exceeded the 100 ng l⁻¹ MAC; the other compound being *pp'* - DDT (April 1989, Six Mile Water) for which there was no known source or approved use within the catchment. Concentrations of pesticides and their derivatives were generally either close to or below the limits of detection.

Table 4   *Pesticides analyses - Lough Neagh rivers: June 1988-March 1991*

Compound	Samples no.	Detection limit ng l⁻¹	Samples in excess of detection limit no.	Maximum concentration ng l⁻¹
Diazinon	60	10	1	110
HCH-β	60	5	15	20
HCH-γ	60	5	21	20
Aldrin	60	5	8	7
Dieldrin	60	5	12	15
Endrin	60	5	8	20
pp' - DDE	60	5	3	10
pp' - TDE	60	5	3	20
op' - DDT	60	5	4	10
pp' - DDT	60	5	10	120
Heptachlor & metabolites	60	5	18	20
Propetamphos	60	10	none	
Dichlorvos	60	10	none	
Malathion	60	10	none	
Parathion	60	10	none	
Parathion-methyl	60	10	none	
Chloryifos-methyl	60	10	none	
Fenchlorphos	60	10	none	
Hexachloro-benzene	60	5	none	
Chlordane	60	5	none	
HCH-α	60	5	none	

4.3 Stream Invertebrates and Water Chemistry.

Monitoring of stream invertebrates in the Mourne study area showed that BMWP biotic scores declined with increasing altitude of the sampling sites, with some of the lowest scores being recorded at sites 1, 2 and 5 which were upstream of any sheep-dips. These sites, in common with most of the other sites had good water quality as evidenced by high dissolved oxygen and low ammonium concentrations (Table 5). The decline could be related to the low conductivity of these headwaters (Figure 2). Two plots of BMWP score vs stream conductivity are presented, one prior to and the other following June 1993; each showing a decline in the BMWP biotic score at conductivities of less than 80 μs cm^{-1}. Within this general relationship, the mean biotic score at site 4 was depressed relative to conductivity in the pre-dipping period (Figure 2a). During this period the four sites with the highest conductivities (sites 9, 11, 12 & 13) each had upstream dips in their catchments. The biotic scores from these sites were high relative to the low conductivity (<80 μs cm^{-1}) sites but were lower than scores from streams 8 and 10 which had no upstream dips but conductivities of close to 80 μs cm^{-1}. However sites 9, 11 & 13 showed evidence of slight organic pollution as exhibited by maximum BOD concentrations in excess of 5 mgO l^{-1}, (Table 5). The lower biotic scores at sites 9, 11, 12 and 13 was less noticeable post June 1993 (Figure 2b).

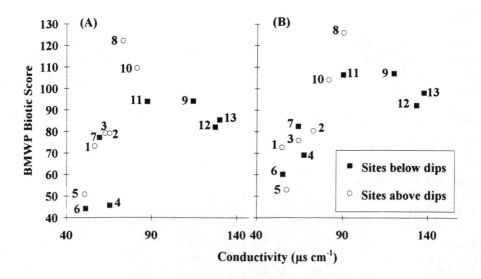

Figure 2 *BMWP biotic score versus conductivity, both (A) before and (B) after the end of compulsory sheep dipping in June 1993.*

Biotic scores from site 4 in 1994 showed a marked improvement relative to those at the upstream site 3 (Figure 3a). That biotic scores were higher at site 4 than at site 3 in 1994 was consistent with the higher conductivity at site 4 (Table 5). The other upstream-downstream pairing, sites 5 and 6, did not show a consistent downstream depression prior to June 1993, although, when comparing autumn samples, the 1994 sample was the first not to show a downstream decrease in biotic score (Figure 3b).

Table 5 *Mourne site physical characteristics and mean values of chemical variables monitored 1991-1994.* [1]*Total soluble phosphorus,* [2]*Soluble reactive phosphorus*

Site	1	2	3	4	5	6	7
Irish Grid Reference	J228254	J225254	J225255	J228266	J243267	J236274	J233278
Altitude (m)	175	170	160	150	180	145	130
DO (% sat.)	97.8	98.7	99.4	100.0	99.7	101.4	101.5
BOD (mg l^{-1})	3.05	1.99	2.18	2.18	1.93	1.95	2.14
Temperature (^{O}C)	8.6	8.6	8.7	8.6	8.6	9.3	9.3
Total phosphorus ($\mu g\ l^{-i}$)	32.7	34.8	44.7	37.0	26.8	25.3	30.8
TSP[1] ($\mu g\ l^{-1}$)	20.4	14.3	16.7	17.3	11.7	11.4	15.3
SRP[2] ($\mu g\ l^{-1}$)	7.24	4.80	4.99	7.79	4.93	4.95	5.94
Ammonium ($\mu g\ l^{-1}$)	18.3	10.7	9.2	18.1	10.3	10.7	13.8
Nitrate ($\mu g\ l^{-1}$)	204.3	147.1	161.4	195.7	114.3	130.0	210.0
Nitrite ($\mu g\ l^{-1}$)	5.9	3.0	3.7	3.8	3.6	3.9	4.8
pH (Geometric)	6.71	6.52	6.51	6.59	6.52	6.48	6.62
pH (Arithmetic)	7.15	6.82	6.80	6.86	6.81	6.74	6.82
Conductivity ($\mu S\ cm^{-1}$)	55.6	69.0	63.5	66.5	54.0	53.2	61.6
Calcium (mg l^{-1})	1.24	1.70	2.34	1.86	1.21	1.30	1.52
Magnesium (mg l^{-1})	0.77	1.00	0.97	1.11	0.87	0.84	0.99
Sodium (mg l^{-1})	8.84	8.32	6.73	7.20	6.4	6.67	6.08
Potassium (mg l^{-1})	0.83	0.81	0.60	0.91	0.59	0.76	0.40

Site	8	9	10	11	12	13
Irish Grid Reference	J240288	J239289	J255284	J218291	J217290	J213276
Altitude (m)	130	130	180	105	105	135
DO (% sat.)	99.1	98.5	98.0	103.1	102.0	99.9
BOD (mg l^{-1})	1.94	2.76	2.05	3.36	2.70	4.50
Temperature (^{O}C)	9.5	9.6	9.3	9.3	9.2	9.0
Total phosphorus ($\mu g\ l^{-i}$)	31.5	83.0	32.5	92.7	86.2	129.5
TSP[1] ($\mu g\ l^{-1}$)	16.7	60.7	19.6	51.1	58.1	76.3
SRP[2] ($\mu g\ l^{-1}$)	6.18	32.43	6.94	21.01	33.67	41.32
Ammonium ($\mu g\ l^{-1}$)	11.5	44.6	12.6	15.5	70.8	22.4
Nitrate ($\mu g\ l^{-1}$)	258.6	767.1	238.6	454.3	528.6	614.3
Nitrite ($\mu g\ l^{-1}$)	5.0	10.7	8.7	10.0	11.0	17.7
pH (Geometric)	6.59	6.81	6.80	6.89	6.98	6.94
pH (Arithmetic)	6.79	7.02	6.94	7.03	7.16	7.18
Conductivity ($\mu S\ cm^{-1}$)	81.6	117.4	81.7	89.0	130.4	133.7
Calcium (mg l^{-1})	3.20	6.96	5.25	4.33	9.44	11.17
Magnesium (mg l^{-1})	1.29	2.29	1.63	1.47	2.58	3.41
Sodium (mg l^{-1})	8.53	7.86	6.69	7.29	8.27	7.83
Potassium (mg l^{-1})	0.87	1.38	0.72	0.79	1.75	2.07

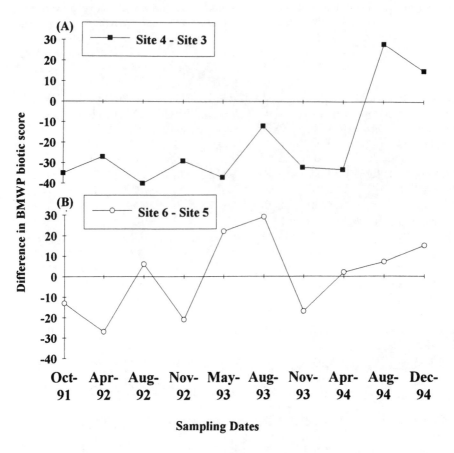

Figure 3 *Differences between BMWP scores at paired sites above and below sheep dips*

Table 6 *Mean BMWP biotic scores at Mourne sites before and after the end of compulsory dipping in June 1993. SD = sample standard deviation.*

BMWP Biotic Scores Above Dips				*BMWP Biotic Scores Below Dips*			
Site	*Before (B)*	*After (A)*	*A - B*	*Site*	*Before (B)*	*After (A)*	*A - B*
1	73.2	72.6	-0.6	4	45.6	69.0	23.4
2	79.0	80.4	1.4	6	44.2	60.0	15.8
3	79.2	75.8	-3.4	7	77.2	82.4	5.2
5	50.8	52.8	2.0	9	94.0	107.0	13.0
8	122.0	126.0	4.0	11	93.8	106.4	12.6
10	109.4	104.0	-5.4	12	82.0	92.2	10.2
				13	85.4	98.0	12.6
Mean	85.60	85.27	-0.33	Mean	72.80	86.17	13.37
SD	25.85	25.84	3.53	SD	22.59	19.36	6.07

Both sites 4 and 6 showed an increase in mean biotic score when samples taken prior to and subsequent to June 1993 were compared (Table 6). Due to the marked degree of variability at individual sites, these increases were not significant at the $p<0.05$ level. However a comparison of biotic scores before and after June 1993, shows increases at each of the sites which had upstream dips, while the sites which were independent of dips showed only small variations in biotic scores (Table 6). Comparing the mean change of the dip dependent group with that of the dip independent group showed that the difference between the two was significant at the $p<0.01$ level, indicating that the removal of the compulsory requirement to dip was associated with an increase in biotic scores.

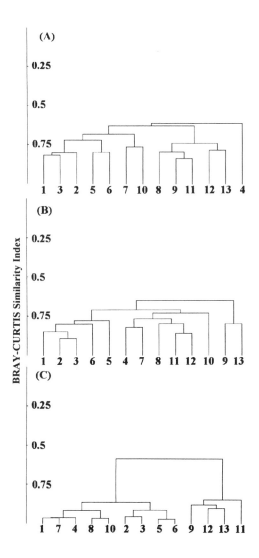

Figure 4   *Cluster analysis dendrograms for: a) biological data pre June 1993; b) biological data post June 1993 and c) chemical data 1991-1994.*

The anomalous invertebrate community at site 4 prior to 1994 is also evident from a comparison of the dendrograms obtained from the cluster analyses (Figure 4). Separation of sites on the dendrogram at low Bray-Curtis similarity index values indicates a high degree of divergence in community structure. For the period up to June 1993, site 4 stream invertebrate community was the first to separate from the dendrogram (Figure 4a). For the subsequent period, site 4 does not diverge until a high Bray-Curtis value is reached and the site is paired with the immediately downstream site 7 (Figure 4b). The remaining dendrogram employed the stream chemistry data of Table 5 and shows a broad division of the sites into two groups which corresponds to a lowland and upland division (Figure 4c). Within each group however there was a high degree of similarity between sites.

DISCUSSION

Between 1987 and 1993 ten water pollution incidents were attributed to sheep dip in Northern Ireland out of 3487 farm pollution incidents, suggesting that incorrect disposal of waste or surplus sheep-dip was not a major problem. However the widespread occurrence of diazinon and propetamphos at concentrations in excess of 100 ng l^{-1} has been reported in many streams and rivers in the Scottish River Tweed catchment during an autumn dip period despite minimal numbers of reported sheep dip pollution incidents.[8] Within the Tweed catchment 5.3% of dips were considered to present a high water pollution risk and monitoring of invertebrates showed severe downstream reductions of BMWP biotic scores in the range 32% to 100% at 15 of 17 high risk sites.[8] This contrasts with the present study which, although it has produced evidence to suggest that sheep-dip disposal was having a limited impact on water quality, did not produce the dramatic results such as those observed in the Tweed. This conclusion may also be drawn from a failure to detect OPs in the Mourne study area during stream monitoring in October and November of 1991.[10]

In the present study interpretation of biotic scores was made difficult due to their inherent variability. Within Great Britain, this problem has been addressed by the RIVPACS computer programme for the prediction of the occurrence of stream invertebrates,[20] but the applicability of this system to the naturally less diverse fauna of Irish streams remains to be fully evaluated. It is apparent however that biotic scores in the Upper Bann streams were depressed in upland areas relative to the lowland streams by factors other than sheep-dip. The relationship identified between biotic scores and conductivity may be a surrogate for other factors, such as nutrient supply and/or acid runoff events which can occur in these poorly buffered streams.

Despite this underlying variability in biotic scores, the 44% depression in biotic scores observed downstream of site 3 is comparable with the impact recorded at high risk sites in the Tweed.[8] The dip downstream of site 3 had two distinct problems which would categorise it as a high risk. The pen, in which sheep stood post-dipping, was beside the stream creating a significant risk of dip draining from fleece entering the stream; the degree of risk depends on ground conditions, weather and numbers of sheep being dipped. In addition, although waste dip from the dip bath drained to a below ground tank, this tank had an overflow pipe which led directly to the stream. Why biotic scores downstream of this site should have improved post 1993 cannot be stated with certainty. Anecdotal evidence from farmers and Department of Agriculture for Northern Ireland veterinary staff suggests that the removal of the compulsory dipping has been followed by a significant reduction in the level of dipping counterbalanced, in some instances, by controlling ecto-parasites by the injection of sheep with acaracides. If this situation is mirrored in the Mourne study area, a reduced level of dipping at this high risk dip would have lowered pollution risk through less direct runoff from the holding pen and reduced probability of the waste storage tank overflowing.

The lowland streams had high biotic scores despite a high density of dips, suggesting that the degree of impact was not great. Additionally these streams suffered from slight pollution from farm organic wastes (slurry and silage effluent) which would depress biotic scores.[11] However as the lowland streams with dips showed a measure of improvement post June 1993, which was not evident in the control sites, it may be assumed that changes in dipping practice which followed the cessation of the compulsory dip had some beneficial impact on water quality.

The general absence of OP compounds in the Lough Neagh major rivers strengthens the impression that sheep dip was not entering rivers in appreciable amounts. By 1988, two non-phenol active ingredients, diazinon and propetamphos dominated use which in theory should have aided the detection of OPs in water. The decline in dip-usage between 1981 and 1988 may reflect change in composition of the dips, as the phenol component declined by a greater proportion than did the non-phenol component. However, a greater emphasis was made in the 1988 survey in allowing for the effects of shared use of dips which is a significant problem as 73.4 % of farms shared dip facilities. Theoretical calculations, which have demonstrated the large volume of water which can be potentially contaminated by a single dip,[21] would suggest that the results of widespread pollution from even a small percentage of dips (1-5%) should have been apparent in the river pesticide monitoring. Additionally, if the sole positive occurrence of diazinon which was observed originated from a dip, the fact that the concentration exceeded the EC Drinking Water Directive MAC for an individual pesticide would confirm the hypothesis that a single pollution event from a dip can have significant consequences with respect to pollution compliance.[21] The results of the river monitoring were also of interest in that they demonstrated a marked reduction in the concentrations of α and γ BHC present. Samples taken in 1974 and 1975 revealed high average BHC concentrations in each of the Lough Neagh rivers. The concentrations were proportional to the human population density in each catchment and the BHC was shown to be entering the rivers via sewage treatment works.[14]

The survey of waste disposal techniques recorded a sharp decline in the number of dip operators which admitted that waste dip entered a stream or ditch, which may reflect a genuine improvement or an increased awareness of the "correct" answer. There is some evidence for the latter explanation as a subsequent survey in 1989-90 involving direct examination of dip sites found that the plug on the dip bath was removed and waste allowed to drain away in 25% of dips.[22] In these instances waste dip must almost always flow initially along a ditch or drain but the ultimate destiny of these drains is in question as often they peter out as they transverse open grassland or moor-land. Thus, waste dip may be said to be disposed to the surrounding land, the most common response recorded the 1988 survey. Many studies have shown that diazinon, propetamphos and other OPs have a high capacity to be stripped out of solution by soils and vegetation where they undergo relatively rapid degradation.[23-26] In addition, repeated soil applications of OPs, including diazinon can lead to enhanced loss rates in the soil, as this facilitates the selection of bacterial populations adapted to breaking down the OPs in question.[27]

Within the Mourne study area, the waste from the dip between sites 5 and 6, was released to a surface drain and proceeded to flow across the hillside grass in an uncontrolled fashion towards a stream some 300 metres away. Soil samples taken between the dip and the stream demonstrated that diazinon and propetamphos the OPs were strongly absorbed by the soil close to the dip site with no evidence that they were present in soils close to the stream.[10]

For this reason it was thought that this dip had a negligible impact on the adjacent stream but the increased biotic scores observed in 1994 at the downstream site 6 suggests that the dip may have been having a limited impact on stream fauna pre-1994, particularly during the autumn dip.

Despite the obvious lack of management control in removing the plug of the dip bath and allowing the waste dip to drain away, the polluting consequences of this type of discharge do appear to be much less than may be first imagined. It is argued that the high affinity of OPs to be strongly bound to soils and vegetation is the principal factor which currently ameliorates these poor disposal techniques evident at many dip sites within Northern Ireland and accounts for the lack of widespread impact observed in surface waters. That the situation within Northern Ireland appears to be better than observed in Scottish catchments may be related to the smaller average flock size in Northern Ireland compared to Scotland,[6,4] leading to lower volumes of dip being disposed of at individual sites which are within the capacity of the surrounding soils to absorb.

ACKNOWLEDGEMENTS

We wish to thank Margaret Gourley and Tim Mackie for their assistance with water and stream invertebrate sampling and analysis.

REFERENCES

1   R. Stephens, A. Spurgen, I. A. Calvert, J. Beach, L. S. Levy, H. Berry,. and J. M. Harrington, *Lancet*, 1995, **345**, 1135.
2   H. B. N. Hynes, *Ann. Trop. Med. & Parasitol.*, 1961, **55**, 192.
3   F. B. Leech and W. D. McCrae, *Pestic. Sci.*, 1970, *1*, 53
4   J. R. Cutler, *Pestic. Sci.*, 1973, **6**, 616.
5   H. M. Bowen, J. R. Cutler and I. R Craigie, *Pestic. Sci.*, 1978, **13**, 563
6   S. Jess, and R. J. Marks, *Rec. Agric. Res. (Dept. Agric., N. Ireland)*, 1986, **34**, 61
7   J. W. Littlejohn and M. A. L. Melvin,. *J. Inst. Wat. Environ. Mgt.*, 1991, **5**, 21.
8   W. A. Virtue and A. M. Church, *J. Inst. Wat. Environ. Mgt.*, 1993, 7, 395.
9   Northern Ireland Annual Farm Census, Department of Agriculture for Northern Ireland, Belfast, 1969-1994.
10 J. Blackmore and L. Clark, 'The effects of the disposal of sheep dip waste on water quality', WRc Reports NR 3067/4244 & 3068/4244, 1993.
11 R. H. Foy and M. Kirk, *J. Inst. Wat. Environ. Mgt*, 1995, **9**,(in press)
12 R. H. Foy, 'Water Quality in Relation to Farming Practises and Capital Grant Availability', Department of Agriculture for Northern Ireland, Belfast, 1991.
13 R. V. Smith, *Wat. Res.*, 1977, **11**, 453.
14 D. B. Harper, R. V. Smith and D. M. Gotto, *Environ. Pollut.*, 177, **12**, 223.
15 S. H. Mitchell and S. Kennedy, *Sci. Total Environ.*, 1992, **115**, 163.
16 R. K. Chesters, 'Biological Monitoring Working Party. The 1978 national testing exercise', Technical Memorandum 19, Department of the Environment, London, 1980.
17 G. W. Snedecor and W. G. Cochran, 'Statistical Methods, 6th edition', Iowa State University Press, Ames, 1971.
18 J. R. Bray and J. T. Curtis, *Ecol. Monographs.* 1957, **27**, 325.
19 C. Sheppard, 'Statistical analyses v.2.1 by scientific software', Marine Environmental Consultants Ltd, 1993.

20 J. F. Wright, M. T. Furse, P. D. Armitage and D. Moss, *Arch. Hydrobiol.*, 1993, **127**, 319-326.

21 J. W. Littlejohn and M. A. L. Melvin, *Environ. Tech. Let.*, 1989, **10**, 1051.

22 T. S. Dick, 'Farm waste and capital grants survey 1989', Department of Agriculture for Northern Ireland, Belfast, 1991.

23 T. D. Inch, R. V. Ley, and D. Utley, *Pestic. Sci.*, 1972, **3**, 243.

24 M. K. Sears, C. Bowhey, H. Braun and G. R. Stephenson, *Pestic. Sci.*, 1987, **20**, 223.

25 D. E. Mullins, R. W. Young, C. P. Palmer, R. H. Hamilton and P. C. Sheretz,. *Pestic. Sci.*, 1989, **25**, 241.

26 F. Bro-Rasmussen, N. Noøddegaard and K. Vodum-Clausen, *Pestic. Sci.*, 1971, **1**,179.

27 M. Forrest, K. A. Lord, and N. Walker,. *Environ. Poll.* (Series A) 1981, **24**, 93.

Pesticides in Freshwaters from Arable Use

P. A. Chave

NATIONAL RIVERS AUTHORITY, RIVERS HOUSE, WATERSIDE DRIVE, AZTEC WEST,
BRISTOL BS12 3UD, UK

1 INTRODUCTION

The Food and Environmental Protection Act 1985,[1] defines pesticides as any substance, preparation or organism prepared or used among other uses to protect plants or wood or other plant products from harmful organisms; to regulate the growth of plants; to give protection against harmful creatures; or to render such creatures harmless. The term pesticide therefore covers a wide range of substances including herbicides, fungicides, insecticides, rodenticides, soil sterilants, wood preservatives and surface biocides.

The common feature of all of these substances is their toxicity to plants and animals of one sort or another.

Because the NRA is the 'Guardian of the Water Environment' it is interested in, and concerned about, anything which may affect the components of that environment - in particular substances which can affect plant and animal life in water.   The Water Resources Act 1991,[2] under which the NRA operates, gives it particular responsibilities. These include a duty to monitor the extent of pollution in controlled waters (defined in the Act as surface and underground freshwaters, and estuarine and coastal waters out to the three mile limit), and powers to take action in the case of polluting inputs being made to such waters so that any quality objectives for the waters are met. The powers include prosecuting persons who cause or knowingly permit polluting matter to enter such waters, powers to take remedial action to deal with pollutions, or to take pre-emptive action to prevent such an occurrence; and powers to give consent to discharges so that they may be controlled in such a way that pollution is avoided.

Pesticides are likely to be toxic to some forms of aquatic life, and must therefore be regarded as polluting materials, and subjected to the most stringent of controls. That this is the case is self evident by examination of the various restrictions and approvals which are required to be observed for their introduction and use. For example, in the UK some six Government departments are involved in the approvals process relying primarily on advice from the Advisory Committee on Pesticides, which examines comprehensive data on toxicity to aquatic and other organisms, bioaccumulation and persistence before recommending the release of the product.   Concern within the European Union over the techniques used to assess the safety of such substances has led to a move to harmonise the procedures through the introduction of a new Directive (91/414/EEC)[3] concerning the placing of plant protection products on the market, in which uniform principles for the authorisation process have been adopted by Member States. Active ingredients are approved at Community level and placed on an approved list. The Directive will only permit authorisation through this route if the pesticide is not expected to occur in groundwater at concentrations above 0.1 micrograms per litre. Some 87 compounds have so far been put forward for review under this Directive.

2 STANDARDS

In order to perform its various functions, the NRA requires standards to use as targets for applying its control provisions, and by which it can judge the success of any such work. As its interests lie in the water environment, the standards are those which should be applied to prevent damage to controlled waters. The standards for pesticides in the water environment which are available for use by the NRA and similar Regulatory bodies are of two types - the statutory requirements set out in European or domestic legislation, and non-statutory quality standards. Both types are commonly referred to as Environmental Quality Standards (EQSs). EQSs are concentrations of substances which must not be exceeded if a specified use of the environment is to be maintained.

The number of statutory EQSs available is presently limited to those derived from EC legislation and transposed by the Government into UK legislation. These arise from the EC Directive on pollution caused by certain dangerous substances to the aquatic environment (76/464/EEC)[4] and its later, more detailed, daughter Directives, which have been transposed into UK law through the Surface Waters (Dangerous Substances) (Classification) Regulations 1989 and 1992.

This Directive places a number of pesticides and related compounds into a List I (the so-called 'black list') and requires Member States to eliminate pollution due to these substances. Biocides not in this list are included in a second list, List II, for which States have to reduce pollution. The Directive, and a series of daughter Directives covering individual substances, lays down emission standards where authorisation of discharges is the appropriate control mechanism, and environmental quality standards where such direct control is inappropriate.

Pesticides also appear in List I and List II of the EC Directive on the protection of groundwater against pollution caused by certain dangerous substances (80/68/EEC), which does not specify standards for individual compounds, but requires action to prevent them entering groundwater in an uncontrolled manner. This Directive supersedes Article 4 of the above Dangerous Substances Directive relating to groundwater.

Eighteen pesticides are included in the list of 36 substances for which reduction targets were agreed by the Government signatories to the final Declaration of the 3rd International Conference on the North Sea in 1990, requiring a 50% reduction over the period 1985 to 1995, and a further 18 pesticides were scheduled for other action to reduce their impact. Rivers are the major source of such materials in the context of the

North Sea, so the NRA has a significant interest in monitoring and applying the control mechanisms available to it to achieve these targets - although much of the control must be related to use and manufacture.

Directive 75/440/EEC[5] concerning the quality of surface water intended for the abstraction of drinking water includes a limit for total pesticides of 0.001 mg/l to 0.005 mg/l depending upon the degree of treatment needed for the water.

Of more direct concern are limits imposed by the EC Directive relating to the quality of water for human consumption (80/778/EEC)[6]. This currently specifies mandatory limits for pesticides and related products, defining these as: insecticides (persistent organochlorine compounds, organophosphorus compounds, carbamates); herbicides; fungicides; and PCBs and PCTs. Limits are 0.1 micrograms/l for individual compounds and 0.5 micrograms/l for total pesticides. The standards are thus very wide ranging in scope, and very stringent. Despite negotiations to update the Directive (which was first negotiated in the 1970's) these levels remain intact at the present time.

Water supply companies are, of course, very concerned to ensure that water supplied for drinking purposes meets fully the legislative requirements of the Directive and beyond this, that water is wholesome in terms of the Water Industry Act 1991[7]. Advice on this is given in 'Guidance on the Safeguarding of Water Supplies' issued by the Department of the Environment as an adjunct to the 'Water Supply (Water Quality) Regulations 1989 et sec under which Water Undertakers are obliged to meet the requirements of the drinking water Directive and the legal requirements of the Water Industry Act. Where exceedances of the pesticide limits are detected by the Companies, they are obliged to inform the NRA for it to take whatever action is available to it to remedy the situation in respect of the raw water source, and the NRA has an internal protocol for attending to such reports. The Guidance contains advisory limits for some 47 substances which may be used as pesticides.

This tranche of existing legislation gives a basic requirement from the regulatory point of view, but for pesticides where no legal standards currently exist, it is necessary to be able to devise suitable EQSs in order that the NRA can take decisions on giving consents to dischargers who may have traces of the material in their discharges, and in order to be able to decide whether enforcement action is needed following the entry of a pesticide, perhaps inadvertently, into a watercourse.

In order to do this, the Department of the Environment and the NRA have drawn up a protocol which takes into account the toxicity, persistence and bioaccumulation factors for individual pesticides, in order to arrive at a suitable EQS value. The protocol uses a basic premise that there is a certain acceptable concentration of each pollutant which does not produce unacceptable effects on the environment and its uses. The environment has therefore a certain capacity to accommodate pollutants and this can be quantified. The protocol employs a use - based quality objective approach and is aimed at determining the concentration of a substance which must not be exceeded if the specified use of the water is to be maintained.

As a result of these factors, a number of EQSs are now in force, or proposed. These are set out in Table 1.

Table 1: Environmental Quality Standards for Pesticides

Pesticides	Freshwater		
	Maximum ng/l	Annual Average ng/l	95 Percentile
HCH		100	
pp DDT		10	
Total DDT		25	
Pentachlorophenol		2000	
Total Drins		20	
Endrin		5	
Hexachlorobenzene		30	
Total Atrazine/Simazine	10000 (P)	2000 (P)	
Azinphos Methyl	40 (P)	10 (P)	
Dichlorvos		1 (P)	
Endosulfan	300 (P)	3 (P)	
Fenitrothion	250 (P)	10 (P)	
Malathion	500 (P)	10 (P)	
Trifluralin	20000 (P)	100 (P)	
Diazinon	100 (P NRA)	10 (P NRA)	
PCSDs			50
Cyfluthrin			1
Sulcofuron			25000
Flucofuron			1000
Permethrin			10

P = proposed   P NRA = proposed by the NRA

3   PESTICIDE USE

Information of pesticide use is important in the context of anticipation of problems and precautionary work aimed at preventing pollution, and for the design of monitoring schemes both sampling programmes to locate possible pollutions, and in ensuring that analytical capability is made available and methods are developed.

The Ministry of Agriculture, Fisheries and Food (MAFF) publishes annual statistics of pesticide use. Information on the 'Top 50' pesticides is available in terms of tonnage and total area to which they are applied.

Table 2: Usage of Pesticides Above 200 Tonnes Per Annum (Courtesy of Pesticide Usage Survey Group, Harpendon)

Active Ingredient	Tonne/year	Use
Sulphuric Acid	6,023	Leaf desiccant
Isoproturon	2,750	Cereals
Chloromequat	2,214	Growth regulator
Mancozeb	1,208	Fungicide
Chlorothalonil	936	Broad leaved crops, cereals, fruit & vegetables
Mecoprop	607	Cereals, grass
MCPA	590	Cereals, grass
Chlorotoluron	579	Cereals
Sulphur	535	Fungicide and foliar feed
Fenpropimorph	516	Fungicide
Mecoprop P	513	Cereals, grass
Pendimethalin	498	Herbicide
Maneb	466	Fungicide
Trifluralin	347	Cereals, fruit and vegetables
Glyphosate	288	Herbicide, arable and non-agricultural
Tri-allate	262	Herbicide
Fendropidin	259	Fungicide
Carbendazim	255	Fungicide
Metamitron	247	Herbicide

Table 2 shows pesticides used in total annual quantities greater than 200 tonnes per annum in 1992. The Table also indicates which of these are used for arable farming and are relevant to this paper.

In order to further target the sampling programme and pollution prevention initiatives, the NRA uses this data, together with information on catchments gathered for various purposes by other organisations to make judgements as to where best to expend its resources. Until the present time much use has been made of a commercially available report, FARMSTAT, which takes into account such data as rainfall, soil type, cropping information,and other information to predict concentration in surface and groundwaters. An improved system known as POPPIE (Prediction of Pesticide Pollution in the Environment) is being installed within a new NRA national centre of expertise, the recently established Toxic and Persistent Substances centre (TAPS) to enable improved predictions to be made. This system will use models to link and interact the seasonally dynamic factors relating to usage, land management and weather with the spatially variable factors relating to soil, hydrogeological and hydrological characteristics in order to show, on a geographical information system, changes in pesticide use through time and predictions of expected concentrations, and allow targeting of problem substances.

4 SURVEILLANCE MONITORING

Having considered the standards to which the NRA is required to work, and the means of identifying and prioritising pesticide problems in general, it is necessary to collect real-life data on their occurrence in the waters of concern.

A sampling programme operated by the NRA to monitor the extent of pollution in controlled waters includes analysis of 117 different substances in this category. This represents about a quarter of the 450 pesticides approved for use in the UK. The number of pesticides monitored is restricted by virtue of analytical cost, availability of analytical methods and laboratory capability. It is estimated that the NRA's pesticide analysis programme costs around £3 million per annum, and therefore programmes are targeted at those compounds most likely to be found, or which are of specific interest. The NRA designs such programmes on a local basis although subject to central policy guidance, and whilst always including those pesticides subject to direct regulatory control by way of EC Directives and other statutory needs, the individual programmes take account of information on pesticide usage in the area of sampling together with the above mentioned reports. Information on crop type, seasonality and pesticide sales and storage often give a good indication of those individual compounds which may be of concern in a particular area. A long term programme of sampling at the tidal limits of rivers - started by the NRA's predecessors on behalf of the Department of the Environment, known as the 'Harmonised Monitoring Programme' gives a further set of data on inputs of pesticides from rivers to the sea, and how these have changed with time.

5 PESTICIDE DETECTION

There are two levels of interest to the NRA related to the legislative situation described above. First, where Environmental Quality Standards are in place or proposed for a variety of pesticides clearly it is the NRA's responsibility to obtain data to ensure that these are observed, since levels higher than these would represent a situation in which damage to the water environment was occurring. The second level of interest is the drinking water standard of 0.1 microgram/l for individual pesticides.

Table 3 shows the number of sites where samples indicated a failure of the relevant EQS. Most of these substances are not mainly associated with arable farming, but with sheep dipping (diazons) and non-agricultural use. It is interesting and perhaps worrying to note that four of the substances are not approved for agricultural use and three of these have been banned for many years.

This rate of failure is very small when compared with the total number of tests carried out (for example over 2,000 examinations for diazinon were performed usually at targeted sites where they might be expected to occur).

Table 3: The Number of Sites Failing EQS in 1993

Pesticide	Number of Sites Failing EQS
Diazinon	44
Dichlorvos	12
HCH	10
Total drins	9
Total DDT	5
Pentachlorophenol	3
Azinphos methyl	2
Malathion	2
Total atrazine/simazine	2
pp DDT	2
Fenitrothion	1
Hexachlorobenzene	1

The more difficult standard to achieve is the 0.1 mg/l limit for individual pesticides in waters and for drinking purposes. Table 4 shows 1992 and 1993 data for pesticides when compared with this limit. It is important to note however, that the limit applies to water at the consumers tap and not the raw water in the river or aquifer. Many companies have now installed appropriate treatment plants to remove these small traces of pesticide before distributing the water to the public. The level is of interest to the NRA because it indicates a potential pollution, and therefore needs investigating.

The precise group of pesticides exceeding the limit varies from year to year - due possibly to different weather conditions or crops being grown.

Diuron, mecoprop, MCPA and isoproturon have uses in arable farming although they are better known for horticultural use, the remainder, apart from the specific use of atrazine on maize, are generally non-agricultural.

It is interesting to compare the pesticides which are found in waters with their usage as set out in Table 2. Only isoproturon, mecoprop, carbedazin and MCPA appear in the top twenty. However, 12 of the 20 most commonly used pesticides are not routinely tested for by the NRA.

Table 4: Summary of Pesticides Which Exceed 0.1 mg/l in Controlled Waters in 1992 and 1993

Pesticide	% Samples Exceeding Limit 1992	% Samples Exceeding Limit 1993
Diuron	13	19
Mecoprop	15	17
Atrazine	16	13
Carbendazin	-	10
Bentazone	-	10
2, 4 DCPA	4	8
Simazine	12	8
Eulan	-	8
MCPA	4	7
Isoproturon	9	6

5 ANALYTICAL METHODS DEVELOPMENT

The lack of monitoring data for some of the most commonly used materials reflects the difficulty of developing suitable methods at reasonable cost to meet the stringent detection limits required for this field of analysis. Of the 450 pesticides currently on the MAFF approved list less than half have adequate analytical techniques, although this may improve as manufacturers of new compounds are now obliged to provide a suitable method. A lack of methods is apparent in the fungicide area. Reference to table 2 indicates that 5 of the most used substances are fungicides and these are not presently monitored. This problem is partly due to the perception that fungicides did not present a risk, and therefore the incentive to develop methods was not high- this is changing as predictive models indicate a potential to leach from soils. For those pesticides where analytical procedures are adequate the NRA as a regulator is subject to the requirements of legal challenge, and needs international recognition of its results in the context of Directives and international agreements, and thus has to apply particularly stringent standards to its procedures.

The NRA has six chemical laboratories that process about five million determinations per year. Some seven hundred and seventy thousand of them are pesticides, herbicides or other trace organic determinands. These place particularly high demands on analytical skills because of their low limits of detection, their high accuracy and precision demands and not least, the high throughput of work.

These challenges have been addressed by the application of rigorous quality assurance to analytical expertise. NRA laboratories have to carry out the following procedures in sequence and obtain satisfactory results before using analytical methods.

An analytical system capable of producing results of the required accuracy for the determinand in question must be selected. The method must describe, unambiguously and in sufficient detail, the full analytical procedure.

The total standard deviation of individual results for relevant types over a range of concentrations is estimated, as is recovery for the determinand in the sample matrices likely to be encountered.

An approach to quality control must be established, based on quality control charts, as a continuing check on performance when the system is used for routine monitoring. Any problems indicated by the routine control system is investigated immediately and remedial action taken. Participation in an external interlaboratory quality control scheme such as WRc Aquacheck is essential as an independent check on bias. Any evidence from this participation that analytical errors are larger than the acceptable limits will initiate immediate investigation and remedial action

The NRA's analytical performance requirements are summarised in bias, precision and limit of detection targets. Bias is the difference between the mean of many analytical results and the true mean and for the determindands in question is set as 20% of the true values. Precision is the degree of agreement between repeated measurements on the same sample, expressed as the percentage standard deviation of differences and is 15% for these determinands. Limit of detection is defined as the smallest concentration that can be distinguished at an 0.05 probability level from blank measurements. Relevant limits of detection for atrazine, 2, 4-D, dicamba, isoproturon, mecoprep and simazine are 30, 40 100, 40, 40, and 30 nanograms per litre respectively.

Samples are taken in glass bottles and are refrigerated before analysis. Extraction procedures are liquid/liquid or solid phase procedures, with florisil column clean up sometimes used for permethrin. Calibration solutions are also routed through the entire clean up procedure to ensure that recovery is taken account of in the final result.

All positive results above the designated minimum reporting values are confirmed if the chromatographic detector lacks specificity (for example, electron capture detectors in GC, ultra violet detectors in HPLC and single ion monitoring in GC-MS), in complex matrices or for analytes with known environmental interferents. Mass spectrometry is the commonly used detection system. If the component target ion and qualifier ions are all found at the expected retention time and in the correction ratios then this technique provides a good solution to confirmation. Because of the much improved signal to noise ratio obtained when using selected ions compared to scan mode very low limits of detection are possible. Instrument software may allow sensitivity to be pushed still further by the use of special tuning procedures. For example, by increasing the ionisation time for selected segments of the mass spectrum the abundance of chosen ions can be increased. The most satisfactory means of confirmation is the use of full scan mass spectrometry coupled with library search of spectra using appropriate fit and purity criteria.

Atrazine and simazine are analysed together with other triazines and organophosphorus compounds. The pesticides are extracted into dichloromethane using neutral, then alkaline, conditions. The extract is concentrated then solvent exchanged into hexane, which is cleaned up with silica absorbent prior to gas chromatographic analysis. Environmental samples frequently contain interfering substances that co-elute with triazines and the triazines themselves have common ions in their fragmentation patterns that can cause confusion. Particular care is taken to use a suitable polar column for good separation, ion ratio secondary ion and an internal standard.

The phenoxyacid herbicides mecoprop, 2, 4-D and dicamba are extracted using solid phase cartridges. The extract is evaporated to dryness, redissolved in acetone, reacted with pentafluorobenzyl bromide and re-extracted with hexane/toluene. The extract is then cleaned up, prior to analysis by gas chromatography/mass selective detection, using secondary ions and ion ratios for confirmation.

Isoproturon is extracted with other uron herbicides using solid phase cartridges, followed by elution with methanol. The phenyl urea herbicides are separated and quantified using reverse phase HPLC with scanning ultra violet detection. Correct gradient elution improves resolution and separates the components of interest from interfering co-extractants.

6 POLLUTION FROM ARABLE SOURCES OF PESTICIDES

Pollution from the arable use of pesticides arises through a number of activities. Normal spraying and other forms of application may lead to overshoot when the application is in an area of land close to watercourses. More likely, though, is the unanticipated leaching of the correctly applied material either because of over-application, or adverse weather conditions, or the fact that the particular pesticide was not absorbed as expected, and leaching occurred to a higher degree than allowed for. Use of the POPPIE approach should lead to improved knowledge of this particular aspect, and a considerable amount of research into leachability has already taken place.

A particularly good illustration of such problems occurred in the Isle of Wight during 1994. The local Water Company informed the NRA according to the procedures described earlier that an exceedance of the drinking water standard had been detected in the water withdrawn from the River Yar. On detailed investigation it was concluded that pollution of the Yar was caused by spraying of isoproturon and chlorotoluron on land at the top end of the catchment; that the rates of application were in accordance with the current MAFF codes of practice; there was no evidence of spillage or improper disposal of waste material; and the spraying was carried out with due regard to advice on ground conditions and rainfall. Nevertheless the soil was waterlogged and prone to erosion and this was the principal cause of the problem. Further advice is clearly needed for dealing with such conditions, which were unexceptional in every way.

A further example of problems due to the normal farming practices occurred in Devon, where again as the result of a Water Company detecting a pesticide in its water supplies - in this case atrazine in a borehole supply - the NRA has undertaken work to identify the cause of the problem. Atrazine has been phased out as a herbicide for total vegetation control by organisations such as British Rail and Local Councils, but is still marketed for use in agriculture on a limited number of crops and around farm buildings. In particular atrazine is the main herbicide associated with maize crops. Due to dairy practices and cereal subsidies the area of maize grown in the Otter valley is estimated to have doubled over the last two years. There is no obvious alternative to atrazine for this application, so in order to reduce the long term risks from its use, farmers in the area were approached and persuaded to investigate other compounds, and those in sensitive areas near the abstraction boreholes agreed not to use this material.

As well as pollution from normal usage, the possibility of accidental pollution from incidents is a further area of concern. In its report 'Water Pollution Incidents in England and Wales 1993,[8] the NRA dealt with 2039 substantiated incidents involving chemicals, 3.5% of which involved pesticides - 70 separate incidents. Of the total number only four percent fell into the NRA's Category 1 classification, (that is, major incidents) and pesticides accounted for five percent of these. The reduction in incidents is also therefore a priority task for the NRA.

The disposal of spent or excess pesticide is sometimes a problem. There are well documented cases where disposal many years ago of discarded material has still led to pollutions occurring after drums have corroded away, leaving concentrated material to leach into surface or ground water. Such incidents have caused extensive damage to fish, including, in one incident, the closure and restocking of an entire fish hatchery.[9]

7 SOLUTIONS

The controls on the manufacture, sale and use of pesticides is already extensive, and becoming more so, as explained above. Once in the public domain, the fate of pesticides depends almost entirely upon the way in which they are used, and the care taken to prevent inadvertent contamination by improper storage, use and disposal. A number of Codes of Good Practice have been published by MAFF and others including for example:
- Code of Good Agricultural Practice for the Protection of water;
- Code of Practice for Suppliers of Pesticides to Agriculture,Horticulture and Forestry;
- Code of Practice for the Safe Use of Pesticides on Farms and Holdings;
- Guidelines for the Avoidance Limitation and Disposal of Pesticide Waste on Farms.

The NRA itself has produced a number of guidance notes:
- Agricultural Pesticides and Water;
- Prevention of Pollution of Controlled Waters by Pesticides;
- The Use of Herbicides in or Near Water.

Farm visits are regularly made, and advice offered through this mechanism, and through other outlets such as agricultural shows and lectures at agricultural colleges, in order to bring the message home that such Codes and Guidance should be adhered to, and pointing out the consequences of inadequate preventative measures.

In addition the NRA is promoting the adoption of such preventative measures as the provision of buffer zones next to watercourses to afford a measure of protection, although of course some pesticides will eventually migrate through these. The prevention of soil erosion is also seen as a significant measure to deter pesticide migration from arable usage.

The imaginative use of long term set-aside policies offer yet another opportunity to undertake proactive preventative measures.

The advent of the TAPS centre and the identification of problems through increased use of interactive data will enable the NRA to feed advice on real problems into the approvals mechanisms, as will the development of improved analytical capability, especially for those compounds for which methods and data are not currently available. This will enable better decisions to be taken on restrictions or bans on those compounds the use of which gives rise to insoluble problems in the environment.

References

1. Food and Environmental Protection Act, 1985, HMSO.

2. Water Resources Act, 1991, HMSO.

3. Directive concerning the placing of plant protection products on the market, 91/414/EEC, O L230, 19.8.91.

4. Directive on pollution caused by the discharge of certain dangerous substances to the aquatic environment, 76/464/EEC, OJ L129/23, 18.5.76.

5. Directive concerning the quality of surface water intended for the abstraction of drinking water, 75/440/EEC, OJ L194/26, 25.7.75.

6. Directive relating to the quality of water intended for human consumption, 80/778/EEC, OJ L229. 30.8.80.

7. Water Industry Act, 1991, HMSO.

8. Water Pollution Incidents in England and Wales, 1993, Water Quality Series No 21, NRA September 1994.

9. P A Chave and A E Moore. Aldrin and Dieldrin - Banned But Not Banished? Proceeding of UNEP Conference on Environmental Contamination, London, 1984, pp 136-140.

Acknowledgements

Thanks are due to T Long, A Ferguson and D Tester for assistance. The views expressed are the author's and are not necessarily those of the NRA.

Pesticides in Groundwater

A. D. Carter and A. I. J. Heather

SOIL SURVEY AND LAND RESEARCH CENTRE, CRANFIELD UNIVERSITY, SHARDLOW HALL, SHARDLOW, DERBY DE72 2GN, UK

1 INTRODUCTION

Groundwater is a valuable resource that can be vulnerable to contamination by pesticides from normal agricultural use /approved usages as well as from point sources such as leaks, spills and improper disposal misuse of pesticides. Since the late 1970's it has been recognised that groundwater contamination by pesticides can occur as a result of the approved use of a pesticide under normal conditions.   Detections of DBPC (dibromochloropropane) and aldicarb in USA wells led to general concern and subsequent investigations in the mid-1980's revealed that atrazine, simazine, bentazone, dichloropropane, molinate and phenoxy-herbicides were present in some European locations. Further concerns for groundwater quality in Europe were initiated by the implementation of the EC Drinking Water Directive[1] which stipulated that supplies of drinking water (at the tap) should not contain more than $0.1\mu g/l$ of a single pesticide and $0.5\mu g/l$ total pesticides.   These limits reflect the precautionary approach towards groundwater protection within the EC in that they are based on the analytical limit of detection of contaminant pesticides at the time they were established. A review of the directive in 1993/94 has led to the continuance of the $0.1\mu g/l$ parameter but deletion of the total limit in that it is not possible or practical to comply with it. The review yet again reflects the continuing desire of the EC to provide pesticide free drinking water. In the USA and Canada, maximum limits for pesticides are based on toxicological assessments whilst the World Health Authority (WHO) have made recommendations for guideline values for approximately 60 pesticides[2].   Table 1 lists the UK, Department of the Environment's Advisory Values and the corresponding WHO limits for selected pesticides[3]. The DoE limits are intended to provide water suppliers with a health based advisory value in the event of a $0.1\mu g/l$ exceedance.

There is often a hydrological continuity of groundwater with surface water and contaminants can move from one aquatic zone to another. Sensitive aquatic flora and fauna could be at risk where contamination is transferred from ground to surface water. Surface water is however much more vulnerable to direct contamination from other sources such as urban and field drainage systems or drift/overspray.   There is also the

Table 1 *Selected Advisory and Guideline Values for Pesticides in Drinking Water*

Pesticide	DoE Advisory Values µg/l	WHO Guide Line Values µg/l
Aldrin/dieldrin	0.03	0.03
Chlordane	0.1	0.3
DDT	7	1
Hexachlorobenzene	0.2	0.01
Atrazine	2	2
2,4-D	1000	100
Heptachlor	0.1	0.1
Lindane	3	3
Methoxychlor	30	30

possibility that contaminants in irrigation water could cause phytotoxic effects in sensitive crops.

2 GROUNDWATER CHARACTERISTICS

Groundwater constitutes approx 4% of water in the total hydrologic cycle (sea and oceans comprise 94%) and its volume is greater than that of fresh surface water. In many countries groundwater is of a substantial strategic significance in public water supply, for example, in England and Wales in total, it forms 35% of supply, but in some areas it may be the only source. In the Netherlands groundwater supplies on average 70% of the water supply. Groundwater is traditionally considered to be a clean, unpolluted supply which requires little treatment and is very often mixed with less pure surface water supplies. Once groundwater is contaminated it can be difficult and expensive to remediate since it may not be technically or economically feasible to restore the resource.

The size of the aquifer resource depends on the pore and fracture volume of the saturated zone. The National Rivers Authority (NRA) is the statutory authority responsible for groundwater protection in England and Wales and all groundwaters are controlled waters. It defines major aquifers as those highly permeable strata usually with a known or probable presence of significant fracturing. Minor aquifers do not have a high primary permeability or they have variable permeability which limits their water supply potential[4]. The major aquifers are generally more porous or fissured and have less potential for attenuating contaminated recharge entering the unsaturated zone. In the UK the major aquifers are the chalk, limestone and sandstone formations. The lower yielding aquifers are important for water supply on a more local basis.

The depth of water abstraction can vary according to the nature of the aquifer but is normally between 20-250 metres, though it can be less than 10 metres in the case of local private supplies. The depth to groundwater largely determines the potential travel time of a contaminant though there are cases where preferential movement of pesticides in an aquifer is known to occur in fissured rock systems[5]. The age of groundwater can therefore vary from a few years to thousands of years for deep wells with an overlying sealing layer.

3 PATHWAYS OF CONTAMINATION

Pesticides can reach groundwater by a variety of pathways and Figure 1 summarises potential fate in the aquatic environment.

Figure 1   *Pesticide Movement in the Aquatic Environment*

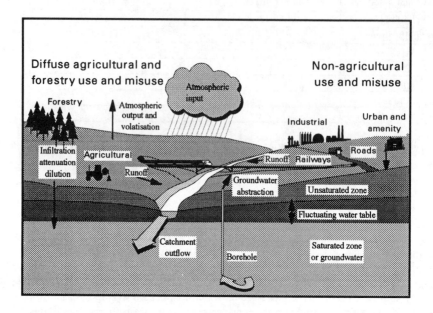

Agricultural applications are normally made to a crop/soil surface whereby there is opportunity for interception, sorption and degradation of the pesticide. The potential travel time to an aquifer is retarded by the soil layers which overlay the aquifer. Other uses of pesticide allow application to surfaces which do not have the retaining layers of soil, for example, those to pavements, railway and road cuttings/embankments. The drainage water passing through these application surfaces can contain pesticide which moves rapidly to underlying groundwater.

There are many incidences of groundwater contamination which are attributed to misuse or accidental spillage of pesticides. Helweg[6] showed that 2,4-D, dichlorprop, parathion and diquat were detected in concentrations of up to 3800μg/l in wells in Denmark. He attributed these high concentrations to direct contamination by farming during filling and rinsing of sprayers and from back-siphoning. Farm soakaways, washing waste and the disposal of sheepdip have contributed to point source contamination of groundwater by pesticides.

4 PESTICIDES DETECTED

The true extent of groundwater contamination is not known since groundwater reserves are so vast, both spatially and in depth and the cost of constructing boreholes for extensive sampling networks is extremely expensive. A report produced for the Dutch government suggested that approximately 65% of groundwater resources in Europe were contaminated by at least one pesticide[7]. The estimates of the report are however based on assumptions that leaching processes and soil hydrologies in Europe are similar to that of the Netherlands which is clearly not the case. Sufficient information has however been collected in some areas to indicate that pesticide transport (as a result of legal approved use) has occurred to some aquifers.

A survey of pesticide contamination of US groundwater up to 1988 discovered 46 compounds in 26 states (Table 2). A USA national well water monitoring survey was commenced in 1988. Results showed that 10.4% of community drinking water wells and 4.2% of domestic wells had at least one pesticide residue detected in their water. A total of 16 pesticide ingredients or metabolites were detected at least once[8]. In California the 1993 update on the Well Inventory Data Base[9] showed the following 13 active ingredients were detected in addition to some of those in Table 2. These were benomyl, dimethoate, endosulfan sulphate, EPTC, ethylene dibromide, methyl bromide, molinate, naphthalene, ortho-dichlorobenzene, prometryn, tetrachloroethylene, thiram and xylene. A total of 112 pesticides or metabolites were monitored for in the Californian programme and simazine was the most frequently detected, followed by atrazine, diethyl-atrazine, diuron, bromacil, bentazone, disopropyl-atrazine, xylene, prometon and 2,3,5,6 TPA. Pesticides were detected in 3.4% ie 80 out of the 2,324 wells sampled.

Table 2   *Pesticides in Groundwater Due to Normal Use in the USA[10].*

Pesticide	States	Pesticide	States	Pesticide	States
1,2-D	4	DDT	3	Methamidophos	1
1,3-D	1	Dacthal	1	Methomyl	1
2,4-D	2	Diazinon	1	Methyl Parathino	1
Alachlor	12	Dicamba	2	Metholachlor	5
Aldicarb	7	Dieldrin	2	Metribuzin	4
Aldrin	2	Dinoseb	3	Oxamyl	3
Arsenic	1	Diuron	1	Parathion	1
Atraton	1	EDB	6	Picloram	3
Atrazine	13	Endosulfan	1	Prometon	1
BHC	1	Ethoprop	1	Propazine	2
Bromacil	2	Fonofos	2	Simazine	7
Carbofuran	3	Hexazinone	1	Sulprofos	1
Chlordane	1	Lindane	3	TDE	1
Chlorothalonil	2	Linuron	1	Toxaphene	1
Cyanazine	6	Malathion	1	Triflurin	4
DBCP	2				

No similar European scale monitoring programme is in place and therefore data on pesticide contamination of groundwaters for selected countries have been compiled in

Table 3 from various sources. The list is by no means comprehensive but provides an indication of active ingredients and metabolites which have been detected. Data have been compiled from van Haasteren,[10] Dobbs *et al*[11] and the National Rivers Authority, UK (Eke, pers comm.).

In England and Wales a national database for detection of pesticides is now compiled from the different NRA regions by the Toxic and Persistant Substances (TAPS) Centre at Peterborough. A total of 18 pesticides or metabolites were found to exceed 0.1µg/l in groundwater in 1993 which were bentazone (14.7), atrazine (11.3), trietazine (4.8), diuron (4.7), pentachlorophenol (3.9), 2,3,6 TBA (3.7), linuron (3.5), clopyralid (3.3), ethofumesate (3.2), isoproturon (2.8), chlorotoluron (2.3), terbutryn (2.2), simazine (2.0), mecoprop (1.5), DDT pp (<1.0), DDT op (<1.0), TDE pp(<1.0) and gamma HCH (<1.0). Pesticides are listed in order of the percentage of samples exceeding the 0.1µg/l limit, though the total number of samples analysed for each pesticide varies *eg* 34 for bentazone and 603 for atrazine.

Table 3   *Pesticide Detections in Groundwater in Selected European Countries*

Pesticide/Metabolite	Countries detected
1,2-dichloropropene	NL
1,3-dichloropropene	NL
2,3,6-TBA	UK
2,4-D	AU, DK, UK
2,4-dichlorophenol	DK
2,6 dichlorobenzamide	NL,DK
Alachlor	AU, I, NL
Aldicarb	NL
Aldrin	UK
Atrazine	AU, DK, F, NL, SW, SUI, UK
Bentazone	NL, SW, UK
Bromacil	NL, SUI
Bromoxynil	DK, UK
Carbetamide	UK
Chloridazone	SUI
Chlorothiamide	NL
Chlorotoluron	UK
Clopyralid	SW, UK
Cyanazine	AU, NL
DDT	UK
Desethylatrazine	SUI
Dicamba	UK
Dichlobenil	NL
Dichlorprop	DK, SW, UK
Dieldrin	UK
Difenzoquat	UK

Table 3 *continued*

Dimethoate	DK, UK
Diquat	DK
Diuron	NL, UK
Endrin	UK
EPTC	UK
Ethoprofos	NL
ETU	NL
Gamma-HCH	AU, NL, UK
HCH	UK
Hexachlorobenzene	UK
Hexazinone	DK
Ioxynil	DK
Isoproturon	DK, SUI, UK
Linuron	UK
Malathio	UK
Mancozeb	NL
Maneb	NL
MCPA	DK, SW, UK
MCPB	UK
Mecoprop	DK, NL, SW, UK
Metalaxyl	NL
Metam sodium	NL
Metamitron	NL
Metolachlor	AU, NL, SUI
MIT	NL
Oxamyl	NL
Paraquat	DK
Parathion	DK
Pentachlorophenol	UK
Permethrin	UK
Pirimiphos	UK
Prometryne	SUI, UK
Propazine	AU, I, UK
Propyzamide	UK
Simazine	AU, DK, F, I, NL, SUI, UK
Tecnazene	UK
Terbutryn	NL, SUI, UK
Terbutylazine	AU, I, SW, SUI
Triallate	UK
Trietazine	UK
Vinclozoline	AU
Zineb	NL

Key : AU Austria, SW Sweden, NL Netherlands, SUI Switzerland, DK Denmark, F France, I Italy.

5 PESTICIDE PROPERTIES WHICH DETERMINE MOVEMENT TO WATER

The physicochemical properties of pesticides which determine their potential fate in the environment are described by Cheng[12], Arnold and Briggs[13]. Hollis[14] describes a classification system based on sorption to organic carbon (Koc) and persistence in soil (half life). This simple classification system allows, for most pesticides, a rapid assessment of a chemical's potential to contaminate groundwater. Gustafson[15] devised a Groundwater Ubiquity Score (GUS), an empirical algorithm, based on these parameters and he was able to show that many of the contaminants in Californian groundwater were expected to leach.

The sorption and persistence characteristics[16], of some of the pesticides in Table 2 suggest that not all detections are attributable to normal approved use *eg* some products are classified as non-mobile and would not be expected to leach and enter groundwater. Those pesticides which do reach groundwater (when used as per label recommendation) have a wide range of mobility and persistence properties and are not necessarily those which would be expected to occur *ie* the most persistent or mobile in the water environment. For example, paraquat and diquat have large sorption coefficients and bind strongly to the soil. They would not be expected to be transported through the unsaturated zone and then detected in free water. Figures 2 and 3 plot the sorption coefficients (Koc) and soil half lives (DT50) of many of the pesticides listed in Tables 2 and 3. Both figures show that there is a wide spread of properties and it is not just the mobile and persistant products which have been detected. This emphasises the complexity of the interactions between a great number of factors and processes which influence the fate of a pesticide when applied to land[17].

Interpretation would be assisted by information on tonnage and hectarage applied and frequency of detection. Further analysis has therefore been carried out on the pesticides detected in groundwater as reported by the NRA for 1992 and 1993. Data on the estimated amount applied (t) and the estimated area treated (ha) on all arable crops treated in 1992 for Great Britain were been obtained from the Pesticide Usage Survey Group[18] and M Thomas (pers comm). (The usage survey has not recorded the use of 2,3,6 TBA since 1982, and the last recorded use of DDT was in 1987).

These data and pesticide properties of Koc, half life and solubility were compiled and a multiple regression analysis carried out to determine whether a combination of these factors were able to explain the frequency of detections of the pesticides in groundwater. No significant relationship was seen with only 16% of the variation being explained. This suggested that the relationship cannot be explained by a simple statistical relationship or that other factors or processes have more effect on pesticide movement to groundwater. A single regression analysis with the GUS index for each pesticide did however explain 72% of the variation. The authors are carrying out further investigation of these relationships using a range of statistical analyses and will report on findings elsewhere.

Figure 2 *Pesticides detected in European groundwater*

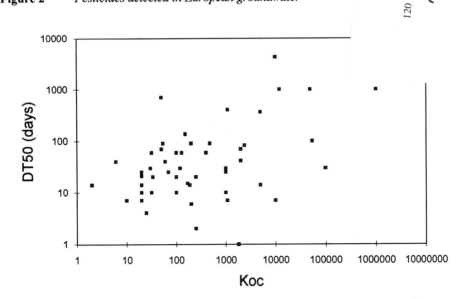

Figure 3 *Pesticides detected in USA Groundwater*

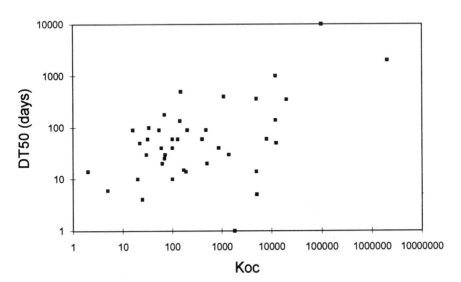

ANALYSIS AND MONITORING

The low levels of pesticide residues present in groundwater have challenged the limits of detection for many of the chemicals and the development of improved analytical methods has become evident as modern data requirements and compliance with the EC directive are imposed on manufacturers. Whilst further surveys have been carried out and a wider range of products analysed for, knowledge on the contamination of groundwater is still very restricted and in many cases uncoordinated. Of the 450 or so pesticides registered in the UK, practical routine analytical methods are only available for approximately one quarter of these. Full compliance with even the revised directive is therefore not possible.

Table 4 *Annex 1A and Red List hazardous substances*

Annex 1A	Red list
HCH *	Gamma HCH **
DDT *	DDT **
Pentachlorophenol *	Pentachlorophenol **
Hexachlorobenzene *	Hexachlorobenzene
Hexachlorobutadiene	Hexachlorobutadiene
Carbontetrachloride	
Chloroform	
Trifluralin	Trifluralin
Endosulfan	Endosulfan
Simazine *	Simazine **
Atrazine *	Atrazine **
Azinophos-ethyl	
Azinphos-methyl	Azinphos-methyl
Fenitrothion	Fenitrothion
Fenthion	
Malathion *	Malathion
Parathion *	
Parathion-methyl	
Dichlorvos	Dichlorvos
Trichloroethylene	
Tetrachloroethylene	
Trichlorbenzene	Trichlorbenzene
1,2-Dichloroethane	1,2-Dichloroethane
Trichloroethane	
Dioxins	
Drins	Aldrin
	Dieldrin
	Endrin
	Polychlorinated Biphenols

Key: * detected in selected European countries (Table 2)
  **detected in groundwater NRA

Water concerns have statutory requirements to monitor for certain pesticides, normally those where there may be a potential concern for human health or those which accumulate in the environment. Those specified by the North Sea agreement (Annex 1A) and the UK's Department of the Environment (Red list) for concern are listed in Table 4. It is now a UK and EC requirement (EC Directive 91/414 concerning the placing of plant protection products on the market) that all new products and those which are being reviewed should have an analytical method with a limit of detection of at least 0.1 µg/l in water.

The cost of a full analysis (*ie* for those products for which a method is available) for a single water sample is very expensive and in the Severn Trent area only 15% of products used in the region are routinely monitored for. In contrast some of the smaller water companies only monitor for the statutory chemicals. The choice of non-statutory products is based on either farm or local authority surveys of pesticide use or an interpretation based on the cropping and the likely products to be used. The development of reliable, cheaper analytical methods is the focus for further research and Eadsforth[19] reviewed some of the newer methods which have the potential to detect to as low as parts per quadrillion. Caution should be used when interpreting any analytical data from water sampling. There are many incidents of false positives being recorded, for example, diuron is routinely monitored for by Severn Trent Water but following further method development all detections prior to 1993 are not now considered to be reliable.

7 GROUNDWATER PROTECTION POLICIES

Protection of water resources from pesticide contamination is a logical, long term option which can be taken to minimise effects and to reduce the overall costs of water treatment. The protection of water resource quality is the responsibility of the NRA in England and Wales who have developed a groundwater protection policy. The policy relies for its implementation on a series of vulnerability maps which will be made available to those involved in all activities which may generate pollution including pesticide usage. The maps and their derivation are discussed further[17].

Hollis[14] describes the development of national aquifer vulnerability assessment scheme which is an integration of soil vulnerability classifications (based on soil hydrology, organic carbon content) with sorption / mobility classifications for pesticides. Soil, aquifer, cropping and climatic parameters are considered to develop a final vulnerability assessment for a particular crop and pesticide. At the catchment level a Geographical Information System has been developed for Severn Trent Water called CatchIS (Catchment information System). Based on similar principals to the above scheme, vulnerability assessments are made using a simple attenuation factor model to predict the likelihood of exceedance of a certain threshold value. The system is under continuous development and will soon contain 1994 cropping data and borehole abstraction recharge zones. Hollis *et al*[20] and Court *et al*[21] describe the system and its potential uses. Severn Trent aim to use CatchIS to identify locations in the catchment which are at risk from contamination and will then target, in consultation with relevant experts, the users who could have the most impact on minimising pesticide contamination of water.

In the USA pesticide and groundwater stategy policies are probably much more advanced than in Europe generally. The strategy coordinates federal, state and private sector roles in addressing the groundwater problem and consists of educational regulatory,

esearch components[8]. The document lists existing problems, then describes its
ɛgy and its implementation. A key part of the strategy is the development of state
ᴍ...ᴀgement plans which are implemented at the local level to address local problems.

In other European countries zoned protection areas have been established around
boreholes eg Switzerland, Sweden, Netherlands, Finland and Germany. Schemes vary but
generally the closest zone is an exclusion area for all pesticides whilst subsequent zones
allow restricted use of pesticides either in quantity applied or type of chemical used.

8 CONCLUSIONS

The EC drinking water directive MAC of 0.1µg/l is a political limit which illustrates the
precautionary principle of 'no pesticides' in groundwater. This limit, together with
concerns over human health, has led to an increase in the sampling frequency and number
of analytes in order to meet the requirements of the directive and public / pressure group
concerns. Many pesticides have been detected in groundwater but the number which are
detected with any frequency are often those which have widespread use and where total
loadings are high. Further improvements or developments are required to provide water
based analytical methodologies for all pesticides since only a quarter of approved pesticides
can be monitored for.

Groundwater contamination from diffuse agricultural pollution is less likely to occur
than from uses whereby the soil zone is by-passed. Misuse and abuse are responsible for
many of the detections in groundwater. Many countries have, or are developing
groundwater or water supply zone protection programmes to minimise or manage the use
of pesticides in vulnerable areas and thus the risk of groundwater contamination.

Acknowledgements

The authors wish to thank the TAPS centre of the National Rivers Authority for providing
data on pesticides in groundwater and the Pesticide Usage Survey Group, CSL, Harpenden
for providing data on application area and tonnage applied.

References

1.   EC Directive relating to The Quality of Water Intended for Human Consumption.
      80/778/EEC, Official Journal of the European Communities, **23**,L229/11
      30 August 1980.
2.   World Health Organisation, 'Guidelines for Drinking Water Quality, 2nd Edition,
      Tables of Guideline Values', WHO, Geneva. 1993
3.   Parliamentary Office on Science and Technology 'Drinking Water Quality,
      Balancing Safety and Costs'. POST, London. 1993
4.   National Rivers Authority 'Policy and Practice for the Protection of Groundwater'
      NRA, Bristol. 1992.
5.   W F Ritter, 'Pesticide Contamination of Ground Water - A Review', American
      Society of Agricultural Engineers, Paper 86-2028, Michigan. 1986

6.   A Helweg In: 'Behaviour of pesticides in water catchmens: modelling and land use practises', Eds, A J Dobbs, T R Roberts. Proceedings of a workshop meeting, August 1991. International Union of Pure and Applied Chemistry, Hamburg. 1991

7.   RIVM/RIZA 'Sustainable use of Groundwater: Problems and Threats in the European Communities'. Ministersseminar Groundwater, Den Haag, 26-27 November 1991.

8.   United States Environmental Protection Agency 'Pesticidesand Ground-Water Strategy'. United States Environmental Protection Agency, Washington DC, USA 1991

9.   Well Water, 'Well Inventory Data Base'. Department of Pesticide Regulation, California. 1993 Update

10.   J A van Haasteren, 'Pesticides and Groundwater'. Council of Europe Press, Strasbourg, 1993.

11.   A J Dobbs, T R Roberts (Eds) 'Behaviour of pesticides in water catchments: modelling and land use practises'. Proceedings of a workshop meeting, August 1991. International Union of Pure and Applied Chemistry, Hamburg. 1991

12.   H H Cheng, 'Pesticides in the Soil Environment: Processes, Impacts and Modeling', Soil Science Society of America, Madison, USA, 1990

13.   D J Arnold, and G G Briggs In: Environmental Fate of Pesticides, Ed, D H Hutson and T R Roberts, 101-122, J Wiley & Sons, Chichester. 1990

14.   J M Hollis In: Pesticide in Soils and Waters: Current Perspectives. Ed,' A Walker, BCPC Monograph, **47**,165-174. 1991.

15.   D I Gustafson. Environmental Toxicology and Chemistry, **8**,339-357. 1989

16.   R D Wauchope, T M Buttler, A G Hornsby, P W M Augustijn-Beckers and J P Burt, *Reviews of Environmental Contamination and Toxicology*, **123**, 1-164. 1992

17.   A D Carter. *Aspects of Applied Biology.***29** 17-23. 1992

18.   Ministry of Agriculture Fisheries and Food, 'Pesticide Usage Survey Report 108: Arable Farm Crops in Great Britain 1992', HMSO, London. 1993.

19.   C V Eadsforth. Brighton Crop Protection Conference - Weeds, 1299-1308. 1993.

20   H M Hollis, C A Keay, S H Gibbons In: Pesticide Movement to Water, Ed: A Walker, BCPC, Farnham, **62**, 345-350. 1995.

21   A C Court, R A Breach, M J Porter In: Pesticide Moverment to Water, Ed: A Walker, BCPC, Farnham, **62**, 381-388.

The Removal of Pesticides During Drinking Water Treatment

B. T. Croll

ANGLIAN WATER SERVICES LIMITED, COMPASS HOUSE, CHIVERS WAY, HISTON, CAMBRIDGE CB4 4ZY, UK

1. INTRODUCTION

The EC Drinking Water Directive (1) brought the subject of pesticides in water to the attention of the public and the media, due to the very low maximum acceptable concentration (MAC) of 0.1ug/l allowed for any individual pesticide and the inclusion of a MAC of 0.5ug/l for total pesticides. These MAC's are not based on toxicological data. The provisions of the Directive have been included in the Water Supply (Water Quality) Regulations 1989 (2) under which drinking water suppliers in the United Kingdom are required to operate. The DoE Drinking Water Inspectorate has issued guidance on the interpretation of the Regulations (3) which includes a list of toxicologically-derived maximum concentrations for UK drinking water (Table 1). Exceedence of these toxicologically-derived numbers will normally lead to closure of the supply immediately. Achievement of the 0.1 and 0.5µg/l MAC's will be within a timescale agreed with the Commission of the EC, for Anglian Water the work will be completed by the end of 1996 with a substantial part already in place.

Table 1. Toxicologically-derived pesticide maximum concentrations for drinking water

Pesticide	Toxicologically derived MAC ug/l
Lindane	3
Dimethoate	3
Mecoprop	10
Atrazine	2
Simazine	10
Chlortoluron	80
Isoproturon	4

The source waters, which are treated to make drinking water, only very rarely exceed the toxicologically-derived maximum concentrations for the treated water as illustrated by the figures from Anglian Water (Table 2). The Anglian Water region is one of intensive arable agriculture and has the highest overall pesticide concentrations in the UK. It will be seen that in surface waters the herbicides Atrazine, Simazine, Isoproturon and Mecoprop were frequently detected and the maximum concentrations were well above above 0.1µg/l. At some sites concentrations were normally above 0.1µg/l. Other pesticides were also detected but less frequently. Only Atrazine was detected in groundwaters, apart from specific pollution incidents, and at few sites and generally below or close to 0.1µg/l (4). After 1989 pesticide concentrations dropped and remained lower than previously, probably due to the extended winter drought, until the last winter (1994/5) when concentrations of Isoproturon similar to those found in the late 1980's have been recorded. In addition, the UK ban on the use of Atrazine and Simazine for total weed control (industrial sites, roads, railways, etc.) has led to a drop in their concentrations but a rise in the detection rate of Diuron which has been used extensively as a replacement.

Table 2   Pesticides Detected Regularly in Surface Waters in Anglian Water Services 1985 - 1989

Pesticide	Concentration		Occurrence (% of samples)
	Normal range (µg/l)	Max (µg/l)	
Lindane	<0.010 to 0.025	0.055	16
Dimethoate	<0.02 to 0.2	0.94	14
Mecoprop	<0.10 to 0.4	5.1	35
Atrazine	<0.02 to 0.6	9.0	58
Simazine	<0.02 to 0.6	7.1	42
Chlortoluron	<0.1 to 0.3	2.6	38
Isoproturon	<0.05 to 1.0	11.5	84

The most extensive forms of "conventional" water treatment consist of :-

> Chlorination
> Chemical coagulation
> Settlement
> Sand filtration
> Disinfection (chlorine)

These treatments remove little of the pesticides detected in UK surface waters and the development of cost-effective methods for their removal to below 0.1µg/l has been the subject of much research over the past decade. The available processes are discussed in this paper.

2. POWDERED ACTIVATED CARBON

Powdered Activated Carbon (PAC) has been used for many years at some surface water treatment sites for the adsorption of naturally-occuring organic compounds which give rise to musty-earthy tastes and odours in the water. Where these problems are not too severe PAC is a cost- effective way of controlling them. Similar considerations apply to the use of PAC for adsorbing pesticides from water i.e. if the pesticide is well adsorbed onto activated carbon and either the occurrence is short-lived or its concentration is close to 0.1ug/l, then PAC can be a cost-effective treatment. The powder is normally added at up to about 20 mg/l either at the chemical coagulation stage or immediately prior to filtration. By adding 10 mg/l at the coagulation stage WRc(5) were able to reduce 0.2µg/l of Atrazine and 0.14µg/l Simazine by 70%. The effectiveness was enhanced by long contact times and addition prior to settlement in an upward flow floc blanket clarifier. Both of these procedures gave a maximum contact time in the available plant (approx. 2hrs) and a build up of PAC in the floc blanket. At a surface water site in Anglian Water PAC doses were much higher to achieve 0.1 µg/l in the treated water owing to the higher pesticide concentrations and it is thought the higher concentration of other organics competing for adsorption sites. At this site granular activated carbon (GAC) filters were more cost-effective.

PAC is not an easy material to handle and inevitably both plant and operators become covered in black dust, this has reduced its popularity. A further snag is the need to control powder addition. Continuous pesticide monitoring is not yet a practical proposition in waters and as their concentration can vary in the water being treated, unless an uneconomically high PAC dose is maintained continuously, there will inevitably be times when the dose is inadequate as concentrations rise. As the MAC is exactly that, i.e. it must never be exceeded, intermittent exceedence is not acceptable. Even if continuous pesticide monitoring were possible the influence of other competing organics could not be accounted for. The modern tendency has therefore been to move away from PAC addition to GAC filters/adsorbers unless the economic arguments were very persuasive.

3. GRANULAR ACTIVATED CARBON

Although charcoal has been used since ancient times for the purification of water, mainly for taste removal, the development of modern GAC filters and adsorbers for large scale water treatment began in the 1960's with the first UK installations in the middle of the decade. However, only two installations were made using very short empty bed contact times (EBCT) for the removal of musty-earthy tastes and odours during the treatment of eutrophic surface waters until the mid 1970's when some of the newly formed Water Authorities in England and Wales began to install experimental filters in order to improve taste and odour control. This was the case in AW and the first full works installations were made in 1984 with all surface water treatment works completed by 1989. These were achieved by replacing the sand in rapid gravity sand filters with GAC of the same particle size. This policy was extremely successful and reduced musty-earthy taste complaints from being the major quality complaint to only a few a year. Although not designed to remove pesticides, these GAC filters obviously did and a summary of their performance on reservior waters, where concentrations are relatively constant and only change slowly is given in table 3.

Table 3 GAC Bedlife for Atrazine to 0.1ug/l Breakthrough (Reservoir Waters).

GAC Type	EBCT mins	Raw Water Mean µg/l	Bed Life Months
TL830	17	0.22	13
F200	19	0.16	24
F200	15	0.12	24
TL830	12	0.29	6
TL830	15	0.27	20
TL830	20	0.28	10

It will be seen that control was achieved to below 0.1µg/l for a period varying between 6 and 24 months. Performance is dependant on GAC type, contact time and the composition of the water being treated. Thus filter sand replacement, which is the cheapest method of GAC installation, can be an effective method of control. However, GAC removal for regeneration or replacement is not easy and, for the shorter bedlives of 6 months to a year, specially designed GAC adsorbers following sand filters are preferred in order to ease GAC handling. Such adsorbers have other process advantages and their installation will normally be determined by a cost-benefit analysis of the whole process relative to the performance required.

Where a river water is being treated with only short storage (1 to 14 days), concentrations can vary much more widely and rapidly as illustraed in fig 1. It also illustrates the ineffectiveness of conventional treatment for the removal of Atrazine.

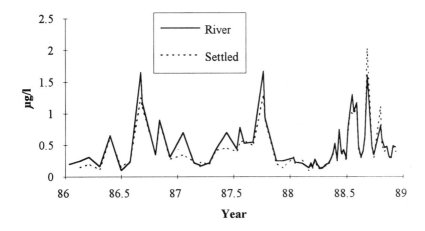

Figure 1. *Atrazine - River and Settled Water Quality*

At the works treating this river, filter sand was replaced by Chemviron F200 GAC (Empty Bed Contact Time (EBCT) of 7.5 mins), its performance for the removal of Atrazine and Simazine is shown in Figure 2 .It will be seen that a regeneration frequency of 4 to 6 months would be needed to control concentrations to below 0.1μg/l. Results for the other herbicides present at the site showed that this frequency was also adequate for their control to below 0.1μg/l. As mentioned earlier, such short regeneration intervals were not considered practical, particularly with filter sand replacement installations, and pilot plant experiments were undertaken to determine the GAC type and contact time required for effective control. The first pilot scale columns indicated that transient breakthrough of herbicides could occur early in the bed-life of the columns. It appeared that the breakthrough coincided with high river water herbicide concentrations. In order to investigate this phenomenon further, columns of 7.5, 15 and 30 mins EBCT were fed with settled water from the works spiked to 2μg/l of herbicides, with the spike raised to 10μg/l for two days a month. The results of this work for 7.5 and 30 mins EBCT are shown in Figure 3 for Atrazine and Chemviron F200 GAC.

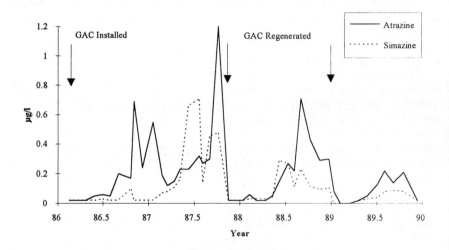

Figure 2. *GAC Sand Replacement Performance*

It will be seen that even at 30 mins EBCT, transient breakthrough above 0.1μg/l can occur early in the bed-life. Interestingly the breakthrough does not occur every month and it is suspected that other water quality factors are causing either displacement of Atrazine from the GAC or its complexation into a form which is poorly absorbed on GAC. Similar results were obtained with Chemviron TL830 GAC for Atrazine and for both GAC's for Isoproturon. More recent results using Chemviron F400 GAC have not so far shown this phenomenen at 15 and 30 mins EBCT but this may be due to the appropriate river water condititons not yet having been encountered.

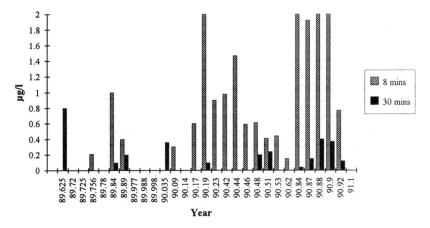

Figure 3. *Atrazine - Transient, Early Breakthrough GAC*

4. OZONE, OZONE + HYDROGEN PEROXIDE

Oxidants have been used extensively in water treatment for disinfection and other purposes such as bleaching of colour caused by the presence of humic materials. Chlorine, ozone and chlorine dioxide have been the most generally effective. All will also oxidise some pesticides but ozone is easily the best for this purpose. It can act either as free ozone or via the production of hydroxide radicals; which is the most effective depends on the pesticide. The latter mechanism can be enhanced by the simultaneous addition of hydrogen peroxide at a ratio of 0.25 to 1 to the ozone dose (in mg/l).

The removal of herbicides by ozone and ozone + hydrogen peroxide has been similar in AW waters to that found by other workers. The results are summarised in Table 4.

Table 4. Herbicide Removal at AW Pilot Plants

Herbicide	% Removal from reservoir water	
	Ozone	**Ozone + H_2O_2**
Atrazine	65 to 78	79 to 85
Mecoprop	75 to 99	86 to 96
Isoproturon	> 90	> 90
Diuron	> 90	> 90

The main disadvantage of oxidants is that they rarely, if ever, degrade the pesticides to carbon dioxide and water and the degradation products may be more toxic than the parent material. Fortunately this has rarely proved the case and even where it has the toxicological limits have not been exceeded. No degradation products as toxic as the parent materials have been detected from the degradation of the herbicides found in UK waters.

5. OZONE + GAC

It will be seen from the previous section that the use of ozone or ozone + hydrogen peroxide upstream of a GAC contactor can therefore extend GAC bed-life, it will also modify the background organics in the river water. It was not known whether this would prevent or exacerbate transient herbicide breakthrough. Further pilot columns have therefore been operated as in section 3 but with ozone prior to GAC. The GAC used was Chemviron F400 and over a period of a year no transient early breakthrough has been observed even with 15 mins GAC EBCT. Thus ozone prior to GAC can extend GAC bed-life considerably and may be economic for pesticide removal, saving both capital and revenue expenditure. These combinations also have advantages for the achievement of other regulatory parameters and improvements in customer service (trihalogenated methanes, bacterial plate counts, discoloured water and nitrite) and have been adopted for use at all but one of AW's surface water treatment works and also by some other water companies for the treatment of eutrophic waters. This treatment is now being installed at 11 AW works, with five being in supply at the present time. The quality improvements at the largest works are shown in Table 5.

Table 5. Full Scale Plant Performance at Grafham

| Parameter | Final Water Quality | | | | | |
| | 1993 | | | 1994 | | |
	Mean	**Min**	**Max**	**Mean**	**Min**	**Max**
Atrazine µg/l	0.07	<0.03	0.110	<0.03	<0.03	<0.03
Simazine µg/l	0.08	<0.03	0.19	<0.03	<0.03	<0.03
Isoproturon µg/l	0.03	<0.02	0.10	<0.02	<0.02	<0.02
Mecoprop µg/l	0.06	<0.02	0.11	0.02	<0.02	0.05
TOC mg/l C	5.2	4.2	8.4	3.0	2.3	3.6
TOC (raw) mg/l C	6.0	4.7	7.3	5.9	5.5	6.7
THM's total µg/l	48.5	1.3	68.0	18.2	8.9	35.2
Iron mg/l	0.06	<0.01	0.49	0.01	<0.01	0.09
Turbidity mg/l	0.31	0.06	1.50	0.12	<0.05	0.36
Colour Hazen	2.8	1.1	4.4	<1.0	<1.0	<1.0
AOC µg/l acetate	52.0			23.0	18.0	30.0
Bromate µg/l				<5	<5	<5

Both ozone and GAC treatment are expensive in water supply terms and current research is aimed at minimising these costs by improving the effectiveness of ozone contactors and determining the maximum GAC bedlife compatible with effective regeneration and acceptably low risk of failing regulatory maximum acceptable concentrations.

6. BIOLOGICAL DEGRADATION

GAC filters and adsorbers are capable of supporting extensive bacteriological growth and in modern waterworks biological action prior to disinfection is encouraged in order to biologically stabilise the water and reduce the growth of non-pathogenic bacteria in distribution systems, where they can lead to enhanced corrosion of pipe materials and aesthetic quality deterioration. This is in contrast to earlier practice where bacterial growths were discouraged by the extensive use of chlorine. Pesticides can be biologically degraded and this is expected to occur during water treatment and particularly in GAC and slow sand filters. Until recently there was little positive proof of this, although it was strongly suspected from results such as those show in Figure 4, where although Atrazine broke through the GAC after about 2 months, it never reached the input concentration prior to regeneration as it should have if only adsorption were occuring.

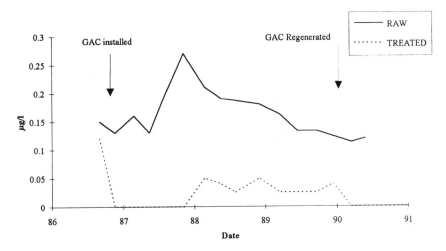

Figure 4. *Atrazine Removal at a Waterworks with GAC Filters*

Recently, collaborative work between AW and the University of Kent has shown that bacteriological degradation of Atrazine does take place in GAC filters and that it should be possible to enhance this effect in a cost effective manner (6).

7. MEMBRANES

Reverse osmosis and nanofiltration membranes are capable of removing pesticides from water. The latter are more attractive because of their lower operating pressures and therefore costs. The removal of some pesticides is shown in Table 6.

Table 6. The removal of Pesticides by membranes

Membrane	Rejection %
Filmtec NF-70	Atrazine 67-89
	Simazine 37-73
Fluid Systems TFCS ULP	Mecoprop 97
	Diuron 97
	Linuron 82
	Isoproturon 97
	Chlortoluron 97
	Atrazine 94

As membranes become cheaper at the same time that conventional treatment is becoming more complex and expensive to achieve tightening standards, it is expected that membrane treatment will be the method of the future particularly as there is pressure to reduce the use of chemicals such as chlorine. However that time is some way off and the critical question of how to dispose of the concentrated reject stream remains to be answered.

One way of using membranes more economically is to combine the use of a low pressure microfiltration membrane (about 0.2μm pore size) with the addition of powdered activated carbon. The use of an Exxflow tubular cloth filter together with PAC for the removal of Atrazine from a groundwater is shown in Figure 5.

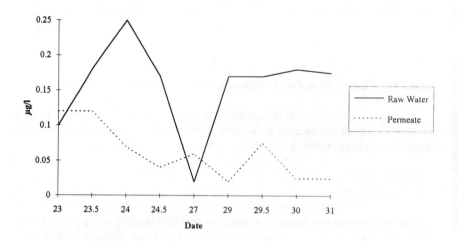

Figure 5. *Atrazine Removal using an Exxflow filter and PAC.*

It will be seen that apart from one break due to plant problems Atrazine was reliably removed after the first day, when the PAC layer on the membrane surface was being built up. This technique avoids high pressures and the used PAC slurry is relatively easy to dispose of. In some situations this technique is now economic.

8. CONCLUSIONS

- Pesticides are present in waters which are treated to produce drinking water at
  concentrations above the Regulatory limits.

-Conventional water treatment processes have a limited capacity for pesticide removal

- Pesticides can be removed to below the Regulatory limits by the use of :-
      Powdered Activated Carbon
      Granular Activated Carbon
      Ozone
      Ozone + hydrogen peroxide
      Biological activity
      Membranes
  or combinations of these processes

- Which process or combination of processes is most cost-effective is site dependent

- The combination of ozone and Granular Activated Carbon filters or adsorbers is at
  present one of the most cost-effective and popular processes due to its ability to
  remove a wide range of pesticides and also achieve other regulatory requirements.

- Membrane treatment is expected to be the process of the future

ACKNOWLEDGEMENTS

The author wishes to thank the Director of Innovation of Anglian Water for permission to publish this paper.

The views in this paper are attributable to the author and do not necessarily represent the views of Anglian Water.

REFERENCES

1.   Council of European Communities (1980). *Directive relating to the quality of water intended for human consumption*. No. 80/778/EEC.

2.   UK Department of the Environment. *The Water Supply (Water Quality) Regulations 1989* Statutory Instrument 1989 No. 1147. Water, England and Wales, HMSO, London

3.   UK department of the Environment. *The water Supply (Water Quality) Regulations 1989. Guidance on Safeguarding the Quality of Public Water Supplies.* DoE Sept 1989.

4.   B. T. Croll, *J Water and Environ. Management,* 1991, **5,** 389

5.   J. Hart and P. R. Carlile, *Pesticide Removal from Lowland Surface Water.* Report No UM-1306 April 1992, The Water Research Centre, Medmenham, Marlow, Bucks., UK.

6.   S. J. Feakin, E. Blackburn and R. G. Burns, *Wat. Res.* 1994, **28,** (11), 2289.

Pesticide Usage in Marine Fish Farming – The Role of the Chemist

Ian M. Davies

SOAFD MARINE LABORATORY, PO BOX 101, VICTORIA ROAD, ABERDEEN AB9 8DB, UK

1 INTRODUCTION

The marine fish farming industry in Scotland has expanded rapidly since the first attempts were made to cultivate Atlantic salmon about 20 years ago. Production has increased from around 5,000 tonnes in 1985 to over 64,000 tonnes in 1994[1]. The industry is concentrated in the sheltered waters of relatively rural parts of the west of the Scottish mainland, the Western Isles, Orkney and Shetland. It now forms a vital part of the economy of these areas, many of which have been revitalised by the advent of fish farming.

Although farmed salmon are harvested from the sea, the cultivation processes also require freshwater resources for the incubation of eggs and for the growing of the young salmon through several juvenile stages (egg, fry, fingerling, parr), until the fish reach the smolt stage at which wild salmon would naturally migrate to the sea. Farmed stocks are also then transferred to sea water for on-growing to the final product. The farmed fish may spend six months to two years in fresh water, and then a similar period in the sea. A few fish are kept in the sea a little longer to allow them to mature sexually, and provide the eggs and sperm needed for the production of further supplies of juveniles.

The role of the chemists in fish farm medicines is to work in a multi-disciplinary partnership with biologists, epidemiologists, toxicologists, hydrographers and environmental scientists to ensure that the right molecules are selected as potential fish medicines, and to bring their particular expertise to bear upon the development of new products and their safe application.

2 USE OF CHEMICALS IN MARINE FISH FARMING

2.1 Antifoulants

There has been a gradual trend in salmon farming, as in all intensive animal rearing, of increasing size of operation, more mechanisation, heavier equipment etc. This has brought about a greater awareness of the vulnerability of stocks of fish kept in enclosed cages to disease, parasites, and adverse environmental quality. Chemical treatments of various kinds have an important role to play in the maintenance of the industry, particularly through the control of fouling organisms, parasites and disease. However, only in the case of antifouling preparations are the substances classified as pesticides. The most notorious group of compounds previously used in antifouling for fish cages are tributyltin compounds. However, the clear indications of adverse environmental effects of these compounds on non-target organisms[2,3,4] led to a prohibition in the UK of their use in aquaculture in 1987. A number of alternative treatments are now available, several of which are based upon copper as the active toxicant.

2.2 Antibiotics

Antibiotics (anti-microbial agents) are used in fish farming to control bacterial diseases. There are a number of compounds in use in the UK (eg oxytetracycline, oxolinic acid, amoxycillin, etc) in a range of commercial formulations. They are almost exclusively administered orally through incorporation into feed pellets. The existing treatments are not always efficacious, and there is an acute need to make available additional compounds. The cost of obtaining authorisation for use in the relatively small aquaculture market is a significant inhibitory factor.

2.3 Parasiticides

An important use of chemicals in fish farming is in the control of parasites. Examples of these chemicals which in other circumstances might be considered to be pesticides, are organophosphorus compounds or pyrethroids. One of the most significant husbandry problems is the effects of infestations of ecto-parasitic salmon lice (*Lepeophtheirus salmonis*), and *Caligus elongatus* on the fish[5]. These small crustaceans are natural parasites of wild salmon (and to a lesser extent some other species). They have a number of life stages in the sea. Pelagic nauplii settle on the salmon and metamorphose through a series of chalimus stages, to the pre-adult and adult stages. All the attached stages feed on the skin and sub-cutaneous tissue of the salmon. In wild salmon caught in fresh water, the presence of salmon lice is an indication that the fish have only recently migrated from the sea into the river. However, the high levels of infestation that have at times developed in fish farms can result in serious damage to the fish and areas of skin and other tissue can be lost, typically from the top of the head. This can result in the fish having difficulty in maintaining osmotic balance, and being more vulnerable to secondary infections[6]. Damaged fish are also of lower value commercially.

The control of salmon lice is essential for any salmon farming enterprise to be successful. At present, the only authorised product available for the control of salmon lice in the UK is Aquagard SLT, the active ingredient of which is the organo-phosphorus compound dichlorvos. The treatment is administered to the fish in the sea. The netting cage containing the fish is reduced in volume, and surrounded by a tarpaulin. Aquagard is added to the water so enclosed to a concentration of 1 ppm of dichlorvos. The fish are treated for one hour, at the end of which the cages are allowed to return to their original size, the tarpaulin is removed, and the treatment chemical is released to the surrounding water. It is necessary to oxygenate the water throughout the treatment, and to observe the fish for signs of undue stress. There is also some degree of risk to operators through exposure to dichlrvos. Protective clothing is necessary, and courses are available providing instruction in safe operating procedures.

3 REGULATION OF FISH MEDICINES

Substances to be used for the treatment of disease and parasites in farmed fish are not controlled under the UK Pesticide Regulations, but are licensed under the Marketing Authorisation for Veterinary Medicinal Products Regulations 1994. Applications for Marketing Authorisation under these Regulations are assessed under three broad headings:

a) Quality, concerning the pharmaceutical quality of the product.
b) Efficacy, concerning the effectiveness of the product in therapeutic use.
c) Safety.   This is a broad heading, including safety to the target species, to the operator, to the consumer, and to the environment. Safety to the operator will not be consider further here, except to note that the product label will always include information on the necessary precautions to be taken. In some cases (eg Aquagard SLT), training courses are available for operators. Likewise, (a) and (b) above will not be addressed in any detail here, although chemists have a central role to play in the manufacturing process and the formulation of effective products.

4 ASSESSMENT OF THE SAFETY OF MEDICINES

4.1 Safety to the Consumer

Safety to the consumer is addressed through the establishment of a Maximum Residue Level (MRL); the level permissible in the fish at the time of harvest. In order to ensure that the MRL is not exceeded, a Withdrawal Period is set within which treated fish may not be harvested for human consumption. The Government, major retail outlets, and the fish farming industry operate monitoring schemes to ensure that fish at retail sale meet all the relevant MRLs.

4.2 Safety to the Target Species

In the case of dichlorvos, safety to the target species (the salmon) presents more of a problem. Dichlorvos is only acutely toxic to the adult and pre-adult life stages of the lice. It is therefore necessary to have a programme of treatment to remove each cohort of juvenile lice as they reach adult stages. The timing of treatments must be carefully assessed by the farmer and his veterinary advisor. If the infestation consists of clearly defined cohorts of distinct ages of lice, then a programme of two or three treatments can effectively eliminate the population. However, populations of very mixed ages can present a more difficult problem, perhaps requiring a longer, or more frequent series of treatments.

Treatment does have some effect on the fish, reducing the activity of the acetylcholine esterase (AchE) enzymes in the nervous system. If the frequency of treatment is such that the fish are not allowed sufficient time to recover before a further treatment is necessary, it is possible to progressively reduce the AchE levels in the fish. The resultant increasing vulnerability of the fish may be combined with increasing resistance of the lice[7]. In some instances in the past this has led to loss of control, and mortalities of stock.

4.3 Safety to the Environment

Safety of the environment is becoming increasingly important in the assessment of both new and existing medicines. The range of phyla present in the sea, and the range of sensitivities of different species, have required a variety of toxicological studies to be carried out, as exemplified by a series of reports on the toxicity of dichlorvos to non-target organisms[8,9,10,11,12]. A more structured and tiered approach to toxicological data requirements is now in the final stages of development, covering the physico-chemical properties of the product, its fate in the environment, and its potential biological effects (Tables 1-3). All applicants for licences must provide the Tier 1 information. The need for Tier 2 and 3 information is assessed on the basis of information from the preceding Tier. Tier 3 data may only be required in relation to the effects and fate of the substance in the environment, and gives the authorising authority scope to request that complex field dispersion trials, or mesocosm experiments be undertaken in some cases.

Table 1 *Physico-chemical Properties*

TIER 1
Molecular weight
UV/vis absorption spectrum
Melting point
Boiling point
Vapour pressure
Solubility in water
Dissociation constant in water
Octanol/water partition coefficient
TIER 2
Sediment/water adsorption coefficient

Table 2 *Fate*

TIER 1
Hydrolysis half-life, pH 5, 7 and 9
Photolysis half-life
TIER 2
Biodegradation mechanism
Half-life in natural sediment-water systems
Bioconcentration tests

Table 3 *Biological Effects*

TIER 1	
Acute toxicity to:	Juvenile fish
	Juvenile crustacean
	Microalga
TIER 2	
Chronic fish and crustacean tests	
Acute toxicity to:	Macrophyte
	Juvenile molluscs
Acute/chronic toxicity to obligate sediment feeders	
TIER 3	

The ecotoxicological information provides a basis for hazard assessment of the particular product under consideration. It is then necessary to determine the likely degree of exposure of non-target organisms, so that the risk posed to the environment can be assessed. Medicines for use in the sea pose particular problems in this regard. Once a chemical is released to the sea, it becomes subject to a number of complex and interacting processes. For example, dichlorvos is initially released near the surface, largely remains in solution in the surface mixed layer and degrades with a half-life of 4-10 days depending upon the temperature[13]. Residual currents will then move the contaminated water away from the farm, while tidal currents move it up and down the axis of the loch. Eddy diffusion processes will also work to disperse the compound in the surface mixed layer, while some of it may be incorporated into sub-surface water masses. The releases of lice treatments are both periodic and irregular, and difficult to predict. However, they mainly occur in semi-enclosed waters such a sea lochs, where natural dilution and dispersion may be less effective than in the open sea. It is therefore important to assess the likely behaviour of the compounds after release to the sea.

The fate of dichlorvos in the sea can effectively be approached through mathematical modelling of various degrees of complexity, describing the movements of water, dissolved substances, and particles in sea lochs. Once models have been constructed, further field experiments are necessary to confirm that the models do reflect the observable behaviour of the compound of interest. Some approaches that have been taken to modelling the dispersion of medicinal treatments in sea lochs, and the validation of the models have been described[14,15].

Mathematical modelling therefore allows prediction of exposure. The models discussed have concerned dissolved substances, but similar approaches can be taken to substances (eg antibiotics) which readily become associated with particles to derive predictions of the distribution of such substances on the sea bed. These predictions can be used in association with the hazard assessment to provide an assessment of the risk posed to the environment by a particular fish medicine. If these assessments indicate that the product can be used safely, and it satisfies the other criteria, a Marketing Authorisation may be granted. However, beyond this is a further level of control on the actual amount that may be used at particular locations. This is exercised through the Control of Pollution Act 1974, which is administered by the River Purification Authorities (RPAs) in Scotland. The additional control exercised by the RPAs should ensure that environmentally safe levels are not exceeded.

5 FUTURE OF CHEMICAL USAGE IN FISH FARMING

5.1 Anti-foulants

These are the only chemicals used in marine fish farming which require to be registered as pesticides. The use of tributyltin compounds is now prohibited, and some farmers seem wary of introducing alternative antifoulants. Development work on new formulations is continuing, and it may be that if the trend towards larger cages continues, the additional effort required to regularly changes fouled nets may make antifoulants more attractive again. The more widespread availability of net-washing machines reduces the drudgery involved in cleaning fouled nets, and thereby reduces the pressure to use antifoulants.

5.2 Antimicrobial Agents

All antimicrobials are considered as medicines. The usage of the more common compounds (oxytetracycline, oxolinic acid, amoxycillin, and potentiated sulphonamide) has been much reduced through the use of vaccines (eg for furunculosis control) and improved husbandry measures, but access to these compounds will continue to be necessary. The increased resistance to the authorised compounds shown by some strains of fish pathogens emphasises the current shortage of authorised antibiotics.

5.3 Sea Lice Control

Much of the preceding text has been concerned with a chemical approach to sea lice control. However, much has been achieved in recent years by other means. Sea Loch Management Agreements are now firmly established and effective in reducing the incidence and severity of disease and parasite problems. Under such agreements, all the fish farm operators in a single loch will voluntarily agree to work together to combat disease. They will normally adopt all in/all out stocking policies that will result in only a single year class being present in the loch at any time, and synchronous fallow periods at all farm sites. Only certified disease-free stock will be introduced, the operators will share information on disease and parasite status of their stocks, and will coordinate the usage of antimicrobials and lice control chemicals. In some areas, the results have been spectacularly good, with improved growth and survival, and reduced usage of medicines.

A biological control strategy based upon the ability of some other species of fish (wrasse) to pick lice off the backs of salmon[16] has seen widespread adoption in Scotland (and created a new industry catching the wild wrasse). Small numbers of wrasse in a cage of salmon can keep lice infestation at a low level, although heavy infestations may not be effectively controlled.

With increasing experience of fish husbandry and biological control, the usage of dichlorvos has fallen to around 15-25% of its 1989 level (P Dobson, pers. comm.), while the production of fish has doubled over the same period. However, it is still necessary to have chemical control strategies available should other measures be insufficient. The Government has indicated that dichlorvos can only remain in use until suitable alternatives have been established. There are a number of other compounds progressing through the Marketing Authorisation procedure. These include another organophosphorus compound (azamethiphos), a synthetic pyrethroid (cypermethrin), hydrogen peroxide, and chitin growth inhibitors.

REFERENCES

1. Anon, Annual Production Survey of Scottish Fish Farms. SOAFD Marine Laboratory, Aberdeen, 36pp, 1995.
2. I. M. Davies and J. C. McKie, *Mar. Poll. Bull.*, 1987, **18**, 405-407.
3. I. M. Davies, J. Drinkwater, J. C. McKie and P. W. Balls, *Proc. Oceans '87 Conference*, 1987, 1477-1481.
4. S. K. Bailey and I. M. Davies, *Mar. Env. Res.*, 1991, **32**, 187-200.
5. R. Wootten, W. Smith and E. A. Needham, *Proc. R. Soc. Edin,*1982, **81B**, 185-197.
6. H. Jonsdottir, J. E. Bron, R. Wootten and J. F. Turnbull, *Salmo salar L. J. Fish Dis.*, 1992, **15**, 521-527.
7. M. W. Jones, C. Sommerville and R. Wootten, *J. Fish. Dis*, 1992, **15**, 197-202.
8. J. G. McHenery and S. W. Forsyth, *Scottish Fisheries Working Paper 8/91*, 1991.
9. J. G. McHenery, D. Saward and D. D. Seaton, *Scottish Fisheries Working Paper 6/90*, 1990a.
10. J. G. McHenery, C. Francis,, A. Matthews, D. Murison and M. Robertson, *Scottish Fisheries Working Paper 7/90*, 1990b.
11. J. G. McHenery, G. E. Linley-Adams and D. C. Moore, *Scottish Fisheries Working Paper 16/91*, 1991a.

12. N. A. Robertson, D. J. Murison, D.C. Moore and J.G. McHenery, *Scottish Fisheries Working Paper 17/91,* 1991b.

13. D. E. Wells, J. N. Robson and D. M. Finlayson, *Scottish Fisheries Working Paper 13/90,* 1990.

14. W. R. Turrell, *Scottish Fisheries Working Paper 16/90,* 1990.

15. I. M. Davies, W. R. Turrell and D. E. Wells, *Scottish Fisheries Working Paper 15/91,*1991.

16. J. W. Treasurer, *"Pathogens of Wild and Farmed Salmonids: Sea Lice"*, Ellis Horwood, Chichester, pp.335-345, 1993.

Assessing Risk Should Not Be a Hazardous Business

T. E. Tooby

PESTICIDES SAFETY DIRECTORATE, MALLARD HOUSE, KINGS POOL, 3 PEASHOLME GREEN, YORK YO1 2PX, UK

1. INTRODUCTION

The regulation of pesticides in the UK is changing and in the future it will be influenced much more by European Union (EU) thinking as well as national issues and continued public concern over the use of such chemicals. It often appears difficult at times to make balanced judgements taking both the benefits of pesticides and the perceived risks following use into consideration. It is the intention of this paper to highlight some of the problems associated with the assessment of the risk of using pesticides and to briefly describe the new procedures adopted by the EU.

In the 19th Bawden Lecture at the Brighton Conference, Professor Spedding[1] raised the issue of the part that public perception plays in influencing change, even when it is ill-founded and erroneous. In particular he said that the extent of the role that pesticides play in producing the abundant choice and variety of produce is not generally appreciated. Graham[2] took a similar line and stated further that the benefits pesticides have bestowed are now in danger of being swamped in a list of perceived disbenefits. Such disbenefits include consumer risk from residues found in treated crops and drinking water.

The chairman of the Advisory Committee on Pesticides, Professor Sir Colin Berry[3], has stated that risks of pesticides poisoning appear far higher in the minds of the public than is actually the case. Indeed, Spedding[1] went on to say that when similar considerations are applied to risk, it is quite understandable that human fears are not simply related to the actual degree of risk. Furthermore, new risks are nearly always taken far more seriously than those older ones more familiar to us. For example, in 1988 there were nine times as many garden accidents with flowerpots as with pesticides[1]. Given this background, it is not surprising that the public in general find risk or risk/benefit analysis, if it is done at all, rather unconvincing.

2. BENEFITS

Finding evidence of the benefits of pesticides has proved very difficult but according to Conway[4] it rests primarily with the avoidance of crop losses both during production and post-harvest. I believe that it can be put more strongly than this. It should be remembered that pesticides have had enormous effects on levels of production[5] by reducing diseases and blemishes and generally raising the quality of produce.

It should be emphasised that when regulating pesticides and carrying out any risk assessment these benefits must be taken into consideration. One other important aspect is profitability for the farmer. The countryside has to be managed and farmers can only afford to do this if they are profitable. Society takes the reliability of good harvests for granted and needs reminding that the industry has been transformed over the last two decades providing a plentiful supply of cheap food[6].

3. RISKS

Pesticides are biologically active compounds designed in the most part to kill target organisms. Unfortunately non-target species within, and adjacent to, treated crops are also affected, often at environmental concentrations at or below those found in the crop. In addition safety must be assured for consumers from residues in treated crops and for operators exposed to the pesticide during application.

Sidall[7] defined safety in the context of Atomic Energy but it can be extrapolated to our purpose.

"Safety is the degree to which temporary ill health or injury, or chronic or permanent ill health or injury, or death, are controlled, avoided, prevented, made less frequent or less probable in a group of people".

Therefore, there is no such thing as **complete safety**. Before we can achieve any control we must be sure that we know what it is that we want to prevent and how we want to take action. For example, Professor Sir Colin Berry[8] stated that the responsibility for the prevention of malaria gives one a totally different perspective on DDT from one which could be held if one is concerned with birds of prey. Furthermore, he put some uses of pesticides into perspective by producing statistics on farm worker's health. 61 people were killed on farms in the UK in 1989, almost all by machinery. No-one died from the use of herbicides. It is often too easy to forget the consequences of negative regulatory action.

Exploring this a little further we are confronted by two terms which are often misused - **hazard and risk**.

The **hazard** is the state of affairs that can lead to harm. Mortality, tumour production or behavioural changes observed in laboratory animals are examples of hazard.

The **risk**, on the other hand, is the probability of that particular event occurring in practice.

Risk estimation is concerned with the identification of the effect and putting it into environmental context. At its simplest it could be the effect, say mortality, and the likely exposure expressed as a Toxicity Exposure Ratio (TER). More often assessments require a higher level of understanding and evaluation than just a ratio.

The **risk/benefit** analysis is the most difficult to determine and is rarely done. Most often need is substituted for this assessment. In environmental terms at present this could be said to be impossible to estimate as there are no economical assessments on the cost of maintaining the environment. However, for future reference, I think that this will have to be done.

There are some pitfalls in any risk assessment. It can be easy to fall into the trap of extrapolating from hazard to risk. One good illustration of this was the story about the product Alar. The problem occurred with effects seen in mice exposed to daminozide, the active substance. These effects were extrapolated directly to human infants. The true critical intake would have been 19,000 litres of apple juice. Nevertheless, the product was withdrawn because of pressures exerted through supermarkets from pressure groups. Apples grown in good faith that season had to be disposed of and at least one grower lost his livelihood[9].

Another problem with risk assessment is distinguishing associations from causations. In February this year the *Grower*[10] ran an article on lindane and the fact that it has been associated with cancer. Two associations were made in the article. Firstly, that the increase in breast cancer in the UK was somehow directly linked with the use of lindane which may or may not be true but the evidence was not compelling. Secondly, that as lindane was approved on a wide range of horticultural crops this was clearly the source of lindane in humans. In fact there are no residues found in such crops and this has been published by the Working Party on Pesticide Residues for a number of years.

4. LEGISLATION

Regulations are needed to provide effective controls on the sale, supply and use of pesticides. In the UK the Pesticides Safety Directorate (PSD) was set up on 1 April 1993 as an Executive Agency of the Ministry of Agriculture Fisheries and Food. The aims of the agency were described in the Corporate Plan as the protection of the health of human beings, creatures and plants, to safeguard the environment and secure safe, efficient and humane methods of pest control.

The legal basis for pesticide registration in Great Britain is contained in the Food and Environment Protection Act 1985, the Control of Pesticides Regulations 1986, and Directive 91/414/EEC (The Directive). The Directive has been implemented through the Plant Protection Product Regulations, 1995. So the EU rules will gradually replace the national ones and for a while the two systems will run in parallel.

The single driving force is The Single Market and the prevention of Trade Barriers. The single market policy is enshrined in the Single European Act 1988. Why, therefore, has it been so necessary to develop a single European regulatory system and is it truly a single system?

The Community is a collection of widely differing regions and states with individual politics; language; customs; climate; geography; economies; philosophies; and agricultural practices. Each State has developed different regulatory policies and each requires approval before a product can be marketed. The duplication and variation of regulatory decisions across the EU can result in barriers to trade. Therefore, a harmonised system was set up to overcome these barriers.

The Directive lays out a two-tier system of regulation. The first tier requires assessment of data on the active substance. The active substance data requirements are given in Annex II of the Directive. At this stage the main objective is to evaluate the hazard and provided that it is acceptable to the Member States, the active substance will be included in Annex I. Authorisation is not granted at this stage. This tier is conducted in Brussels at the Commission and votes are taken at the Standing Committee on Plant Health.

Once the active substance has been included in Annex I, the second tier can proceed. Products containing the active substance on Annex I must be authorised in each Member State and requires a risk assessment using the data requirements and exposure guidance presented in Annex III. Although this two-tier approach has been criticised, leaving the risk assessment to the Member State has a number of benefits in that it can be tailored to individual agricultural practices and environmental concerns thus introducing risk assessment procedures using the most appropriate risk estimates, hazards and exposure data.

The purpose and main structures of the Directive are given in Fig 1. The third point of the purpose is important and lays out clearly that although there should be no barriers to trade, safety has a priority. The key point is that the decision-making process is based on science which will require the need to conduct a risk assessment.

PURPOSE
-*Harmonisation of the registration process across the EU.*
-*Prevention of barriers to trade in plant protection products and in plant products.*
-*Risks to health, ground-water and the environment and human and animal health should take priority over improving plant production.*

STRUCTURE
-*Scientific and technical knowledge is the basis for decision making.*
-*Risk management is a past of the evaluative process.*
-*There is an obligation to ensure that there are real benefits from use.*

Figure 1 *Purpose and structure of Directive 91/414/EEC*

5. REGULATORY PROCEDURES

An agrochemical company wishing to market a plant protection product will need to seek Annex I listing for the active substance in the first instance. This can be achieved through any Member State willing to act as the rapporteur. Other Member States can grant the product a Provisional Authorisation for 3 years under Article 8.1 which provides a derogation pending the outcome of the vote at the Standing Committee on Plant Health and Annex I listing. Once the active substance has been listed in Annex I, products containing that substance can be authorised in each Member State. The process at this stage should become much easier and it is expected that applicants and Member States will make use of the mutual acceptance provisions in the Directive. This does, however, assume that each Member State has the same level of ability and understanding of risk assessment procedures.

Any product already on the market in the EU will be regulated by 'National Rules', that is the legislation already in place in each Member State. Over the next 10 years it is intended that the pesticides already on the market will be evaluated and brought up to modern standards through a very ambitious review programme. I should say that it is unlikely to be achieved within the tight timetable originally proposed.

6. REVIEW PROCEDURES

The aim of the EU review programme is to re-evaluate all of the pesticides already on the market prior to July 1993. The work load was divided across the 15 Member states and over a 10 year period. This programme has now begun with the re-evaluation of the first 90 compounds. The UK and Germany have been selected to carry out a pilot project on 3 chemicals as a training exercise for all 15 Member States.

The timetable has been delayed because of the accession of three new Member States and the scale of the undertaking. The final reports to be prepared by Member States will be completed by October 1996.

A second list is being discussed now and further guidance is being provided for industry and Member States. It can be seen that although the procedures have been designed as a single harmonised system, the fact that each Member State conducts its own assessment will continue to result in a variety of regulatory decisions.

It is the review programme that is identifying problem areas and an opportunity has arisen for Member States and the Commission to meet regularly to discuss specific regulatory problems. Thus procedures are being developed in the light of experience and, most importantly, the criteria to be used in evaluations will be based on this experience and on the developing mutual understanding of the decision-making processes.

In order to help each Member State come to similar decisions 'Uniform Principles' were developed, which became the Annex VI of the Directive.

7. UNIFORM PRINCIPLES

These principles are to be used only after the active substance has been listed on Annex I and are for product related authorisations at Member State level. They provide guidance to the Member state over the evaluation and decision-making steps to use. They rely on a number of steps ending in '*unless*' clauses. These clauses state that no authorisation can be granted **unless** it can be demonstrated that there are no unacceptable environmental effects. As a guidance document, this statement can be regarded as interesting but not helpful. It is very difficult to reach a consensus over what are acceptable effects in the environment so guidance making reference to unacceptable effects without providing any suitable definition will result in much debate between Member States and much anxiety amongst the Agrochemical Industry.

The principles also rely on trigger values based on TERs and exposure estimates to move from one tier of testing to another. These triggers may be too stringent at present but over the years following practical experience, they could be modified.

One of the key areas of this Annex is the section on 'The influence on the environment' in the decision-making chapter and in particular the section on ground-water protection. It provides a useful illustration of the guidance provided to Member States on decision-making. The section opens with the general statement that no authorisation can be granted if the following are not satisfied;

> -*Laboratory and field soil tests demonstrate that the* $DT_{50} < 3$ *months and the* $DT_{90} < 1$ *year;*
> **or**
> -*Laboratory tests on soil demonstrate that no greater than 70% applied residues remain after 100 days;*

unless there is no effect on wildlife and the soil.

Further advice is given stating that contamination should not exceed the Drinking Water standard for an individual pesticide or a toxicological standard. Provided that the toxicological standard is not exceeded, a *Conditional Authorisation* can be granted for 5 years provided it can be demonstrated that risk reduction measures have been put in place to reduce the contamination of ground-water to meet the requirements of the Drinking Water standard. This applies equally to older pesticides and new compounds entering the market. For older pesticides, it would be expected that monitoring data would be used. New compounds would need to be evaluated using mathematical models.

8. PRACTICAL CONSIDERATIONS

Pesticide application is imprecise and residues are rarely confined just to the crop and can be transported away from target areas contaminating important sectors of the environment. A good example of transport away from the target zone can be found with certain herbicides contaminating surface waters and some ground waters. Identifying the nature of the likely risk of environmental harm from these residues is more difficult and has been main purpose of the procedures being developed for ecotoxicological risk assessment. Furthermore, identifying the transport mechanisms is also important when attempting to assess risk and determining the best and most practical means of reducing that risk. Risk reduction measures will vary depending on the agricultural practices currently in use. This will be discussed during the next few years throughout the EU.

In the UK, aerial application, which can be considered as being less precisely targeted, is very tightly controlled and similar measures have been taken in some parts of the USA. However, ground application is regarded as less of a problem. Drift away from target areas has been studied for some time to determine the extent of exposure likely for operators in the field[11]. The fact that drift does occur and could affect adjacent non-target areas demonstrates how imprecise pesticide application is in general. Beyer[12], from the agrochemical company DuPont, recognised this fact in 1991 and stated that much work needed to be done to improve the hydraulic nozzle which had been the mainstay of pesticide application for the past 40 years.

Using an aquatic example once again, the practical consequences of imprecise application and the risk reduction measures necessary can be illustrated. Many arable crops are grown on land adjacent to drainage ditches or larger surface waters. The applied pesticide does not remain entirely on the crop or target weed or organism. Contamination of the soil below the crop and areas adjacent to the crop can occur. The transport mechanisms are overland flow (including erosive losses); inter-flow (horizontal flow above an impermeable layer); drainage; ground-water seepage; and drift. In the UK, the major routes of surface water contamination are drift and drainage (including macropore flow). The drift component has attracted the attention of several groups in the EU including the UK and as a result field studies have generated good data on the decline in drift fall-out with distance from the treated crop. Two factors emerge from these studies. Firstly, drift does occur and secondly relatively simple risk reduction measures are presented. Fig 2 shows that a relatively small untreated strip adjacent to surface water could prevent much of the fall-out contaminating the aquatic environment. If spraying operations were to take place immediately adjacent to the surface water, some contamination would be expected. Although the contamination would not be a simple calculation derived from overspray, the area under the curve, for the situation illustrated, would be about 5% of the applied dose. If an untreated strip of 5 m could be adopted, the decline in drift fall out would be such that only about 0.5% of the pesticide applied to the crop would be expected to fall-out above the water. In fact a similar rate could be expected from an untreated strip of perhaps a width of only 3 m. Even a 1 m strip would reduce the contamination appreciably.

Currently in the UK for very biologically active compounds the strip has been set at 6m. This has caused considerable concern to farmers and growers as a 6m strip can reduce the available cropping land considerably thus reducing the profitability of the farm. The Advisory Committee on Pesticides has requested PSD to review the procedures for using such strips or 'buffer zones' taking the agricultural impact into consideration. In other words, balancing risk against need.

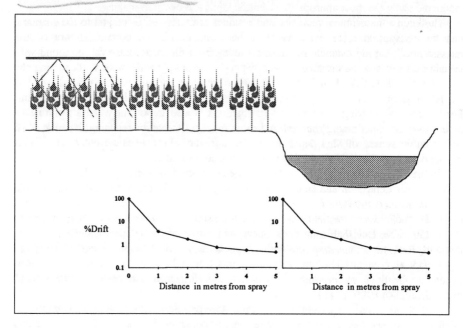

Figure 2 *Decline in drift fall-out with distance*

The environmental impact of spray drift has been little studied. One notable exception has been a study carried out on drift by English Nature in which Pinder *et al*[13] reported that spray drift following ground application of cypermethrin caused mortality of nymphs of *Corixa* sp. and *Notonecta glauca* at distances of up to 15m.

The problem remains that what seems to be a simple risk reduction measure is not. A farmer could not be expected to reduce his growing area without some economic compensation. The unsprayed strip around the crop would be a source of many fungal pathogens and pest species for example. This is fine for some crops but not others. Therefore, this debate will continue over what is the best practice.

Drift is only one of the possible routes for contamination. Drainage probably contributes the greatest to contamination of surface waters. An untreated strip adjacent to water will not affect drainage and other risk reduction measures will need to be taken in such circumstances.

9. CONCLUSIONS

These changes in regulatory procedures have to be taken in the context of a wider initiative by Government in the form of a policy on minimisation - this means the use of pesticides at the minimum effective dose or concentration and not above that level for insurance purposes. Also an important initiative on deregulation has to be considered when conducting risk assessment which means that there should be less not more regulatory hurdles to overcome. All can be achieved provided a good risk assessment has been conducted using the most appropriate scientific input and decision-making.

The key to the problem rests with an assessment of the risk in real terms. The hazard must be relevant and the exposure must be realistic and not compound worst-case assessments. It is only by such means that confidence in the procedure will be maintained, use can continue and risks reduced.

References

1. C. R. W. Spedding, *Proceedings Brighton Crop Protection Conference - Pests and Diseases,* 1992, **1**, 3.
2. J. Graham, 'Food Safety in the human food chain' (Ed. F. A. Miller) CAS Paper 20, Centre for Agricultural Strategy, University of Reading, 1990, 39.
3. C. L. Berry, *Proceedings Brighton Crop Protection Conference - Pests and Diseases,* 1990, **1**, 3.
4. G. R. Conway, 'Pesticide resistance and world food production', Imperial College Centre for Environmental Technology, 48 Princes Gardens, London SW7, 1982.
5. S. Rickard, *Proceedings Brighton Crop Protection Conference - Weeds,* 1991, **2**, 755.
6. W. Griffiths, *Proceedings Brighton Crop Protection Conference - Pests and Diseases,* 1988. **1**. 111.
7. E. Sidall, 'Risk, Fear and Public Safety', Atomic Energy of Canada Ltd, 1980.
8. C. L. Berry, *Trans. Med. Soc. Lond.,* 1992, **108**, 72.
9. E. N. Whelan, 'Toxic Terror', Prometheus Books, Buffalo, New York, 1993.
10. Grower 123, (10), 9 February, 9, 1995.
11. G. A. Lloyd, and G. J. Bell, Report of a study carried out in 1983 in association with the BAA. Subcommittee on Pesticides, SC 7704, 1983.
12. E. M. Beyer, *Proceedings Brighton Crop Protection Conference - Weeds,* 1991, **1**, 3.
13. L. C. V. Pinder, W. A. House and I. S. Farr, 'The environmental effects of pesticide drift' (Ed. A.S. Cooke), English Nature, 1993.

The Control of Pesticides in Water: The EQO/EQS Approach

Melanie G. C. Quinn

WRC PLC, HENLEY ROAD, MEDMENHAM, MARLOW, BUCKINGHAMSHIRE SL7 2HD, UK

1 INTRODUCTION

WRc has, for a number of years, recommended Environmental Quality Standards (EQSs) to the Department of the Environment (DoE), National Rivers Authority (NRA), Scottish River Purification Authorities (RPAs), and DoE Northern Ireland. The DoE requires the adoption of EQSs to meet its obligations to the Dangerous Substance Directive (76/464/EEC). This Directive requires the control of substances (which include pesticides) identified in two lists: List I and List II.

- List I - contains substances thought to be particularly dangerous because of their toxicity, persistence and bioaccumulation; and
- List II - consists of less dangerous substances which nevertheless could have a deleterious effect on the environment.

Under this Directive there are two approaches to controlling List I substances. The first is by setting uniform emission standards (UES) and the second is by using EQSs. For the control of List II substances the Directive requires that National Governments adopt EQSs. The UK favours the EQS approach although for the control of the most dangerous substances, the so-called Red List substances, the UK has adopted a dual approach requiring the application of best available techniques not entailing excessive costs (BATNEEC) and of strict EQSs. The implementation of the EC Dangerous Substances Directive (76/464/EEC) in the UK is shown in Figure 1.

The NRA, RPAs and DoE Northern Ireland similarly use EQSs for the setting of consents to control contaminants in discharges. Thus, the need for EQSs derived by the NRA, RPAs and DoE Northern Ireland is more to resolve localised contamination rather than satisfy legislative requirements.

2 EQO/EQS APPROACH

EQSs are intended to meet agreed Environmental Quality Objectives and can therefore be defined as the concentration that must not be exceeded if the agreed water quality objective or 'use' is not to be compromised. Examples of 'uses' are given in Figure 2.

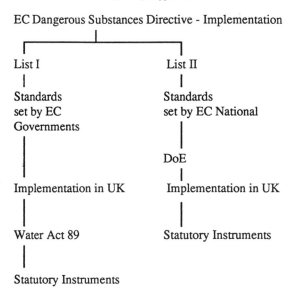

Figure 1 EC Dangerous Substance Directive - Implementation

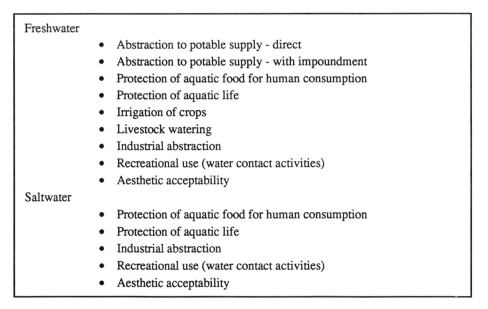

Figure 2 Examples of Identified Water Uses in the UK

However, it is the protection of aquatic life for which the methodology to derive EQSs is most clearly defined and this will be used to illustrate the process in the following sections.

3 DERIVATION OF EQSs FOR PESTICIDES

To date we have derived EQSs for around 60 pesticides. Additionally, we are currently in the process of recommending standards for a further 15 substances.

The process for deriving EQSs is outlined in Figure 3. The dataset required to recommend an EQS is extensive and requires expertise to critically assess both published and un-published data to address the following points:

- What are the sources of contamination of the substance?
- How widespread is its occurrence?
- Can the substance be analysed and at what detection level?
- How will it partition in the environment?
- How persistent is it? What are the degradation products?
- How toxic is the substance to aquatic life?
- Are field data available to compare with laboratory observations?

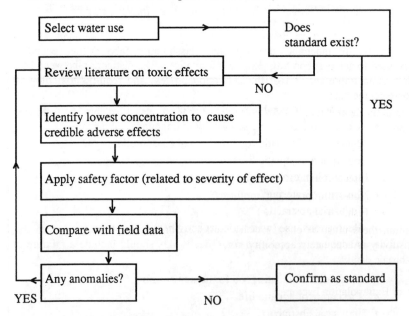

Figure 3 Derivation of Environmental Quality Standards

In order to address the above points, data are assessed for the following areas:
- Chemical and physical properties
- Analysis (e.g. Methodology, limit of detection)
- Behaviour in the environment

— Uses and manufacture
— Entry (e.g. point source and/or diffuse sources)
— Fate (e.g. evaporation, hydrolysis, degradation and adsorption)
- Environmental concentrations (e.g. UK, worldwide)
- Pharmacokinetics, toxicology and bioaccumulation
— Aquatic life
— Mammals (including humans)

3.1   Minimum Aquatic Toxicology Dataset

Before an EQS can be derived for a particular substance a minimum dataset must be satisfied. This is required to balance the requirements for a transparent and consistent approach for the derivation of the standards with the need to allow sufficient flexibility to utilise scientific judgement in the interpretation of all the data available.

The minimum dataset used by WRc takes account of the work carried out on species sensitivity distribution by the EPA and OECD which suggested that the sensitivities of aquatic organisms follow a log normal distribution and that only sets of eight or more data are required to represent "species sensitivity" (i.e. the majority of species, not necessarily the most sensitive ones).

With this in mind, data reported from studies with suitable experimental procedures and test species, recognised toxicity endpoints, and dose-response relationship (i.e. primary data), are considered necessary for the derivation of an EQS. In addition, data from studies not falling under the "primary data" classification (above) (i.e. secondary data) may be used to act as supporting toxicity information. The distinction between both types of data is an important part of the derivation process and requires toxicological expertise to assess the data.

The minimum dataset requires data on species that are classified into the following aquatic groups:

- Algae
- Crustaceans
- Insects (freshwater only)
- Non-Arthropods (e.g. molluscs)
- Fish.

As a minimum there must be eight toxicity values spanning the above groups so that species sensitivity is adequately accounted for. Ideally this should include a mixture of acute and chronic data.

3.2   Use of Extrapolation Factors

The lowest credible adverse effect level is selected from the minimum dataset and an extrapolation factor is applied to this to predict a concentration unlikely to effect aquatic life. Extrapolation factors are needed to account for intra- and inter-species variation; duration of exposure and artificial test conditions. WRc generally adopts the arbitrary safety factors of 100, 10 for acute and chronic data applied to LC/EC_{50} data. These factors can vary if there is a valid scientific reason, such as a substance is highly persistent or is likely to bioaccumulate significantly.

3.3   Comparison with field data

The preliminary EQS, based on laboratory data, is compared with available field data (when available). If there are no anomalies the standard is recommended to the DoE, NRA, RPAs and DoE Northern Ireland. If there are anomalies, these are explored and the EQS may be refined before our final recommendation.

4. DIFFICULTIES AND ADVANTAGES OF THE EQO/EQS APPROACH

The fundamental difficulty of the EQS/EQO approach is that an adequate database is required to derive a standard. Without such a dataset one cannot be confident that the most sensitive species has been identified and, therefore, protected.

Difficulties with implementation include:
- The assumption that all waters of specified types have the same sensitivities, which could lead to over or under protection where standards are derived on an inadequate database (i.e. if the effect of water hardness on the toxicity is not known).
- New dischargers may have more stringent limits than existing dischargers.
- The EQS may be technically unachievable
- and that EQSs require monitoring.

In contrast, there are a number of important advantages with the EQO/EQS approach:
- EQSs are based on scientific data and, therefore, are biologically meaningful;
- EQSs take into account dilution and assimilation capacities of the environment;
- EQSs take into account both diffuse and point sources of contamination.

Storage of Pesticides and the Training of Users

J. C. Seddon

BASIS (REGISTRATION) LIMITED, 2 ST JOHN STREET, ASHBOURNE, DERBYSHIRE DE6 1GH, UK

1 INTRODUCTION

The BASIS registration scheme was established at the behest of the UK government in 1978. It is a practical scheme, which has been tried and tested for fourteen years of distribution and marketing of pesticides.

The scheme was originally established owing to two major fires resulting in pesticide pollution incidents in rivers used for extraction of drinking water. The paper will review the background to the establishment of BASIS; the BASIS Registration Board and administration; BASIS as it relates to current UK legislation; staff training, pesticide storage and transport assessments and progress to date.

2 BACKGROUND TO BASIS

BASIS is an independent registration scheme for distributors, seedsmen, advisers, consultants and contractors of pesticides in the agricultural and non-agricultural pesticides industry. It is concerned with aspects relating to storage and transport of crop protection products and competence of staff involved in storekeeping, sales and advice. It is not a trade association.

In 1978 the Ministry of Agriculture, Fisheries and Food was satisfied with the registration of pesticides. There were two schemes, both of which were non-statutory, looking at the safety and efficacy of pesticides, and these had both been in operation for 40 years. However, Government was concerned about problems with the storage of pesticides and competence of staff involved in sales and storage as a result of two pesticide fires which led to pollution of water supplies.

Members of the trade were called together by the Ministry and it was agreed that the industry would put its own house in order rather than, at that time, have a statutory scheme imposed.

To enable the scheme to work, the manufacturers' trade association agreed only to supply pesticides to people who were registered with BASIS, which gave an incentive for people to register and abide by the standards set. The BASIS scheme and this agreement were cleared by the Office of Fair Trading. However, after a number of years it appeared to contravene the European Commission's Treaty of Rome relating to restrictive practices. It was considered that the BASIS scheme may be a restriction of trade between member states since BASIS registered distributors were only allowed to handle pesticides which had been cleared through the UK Government's Pesticide Safety Precautions Scheme and importation of pesticides from the EU did not meet that criterion. The UK and Ireland were in fact the only two countries within the EU at that time which did not have a statutory pesticides scheme. However, after much deliberation and discussion between Government and the EU it was agreed that legislation should be applied to pesticides in the form of the Food and Environment Protection Act 1985.

During the period between withdrawal of BASIS sanctions by manufacturers and the announcement of pesticide legislation, BASIS continued as a voluntary scheme for a period of some four years, which was obviously a great credit to the pesticide industry.

The aims of the BASIS scheme are: to ensure a high standard of safety in the storage, distribution and contract application of crop protection products; having a due regard for the protection of the environment and to ensure that staff offering advice on such products have received proper training and have achieved proficiency in this field. The scheme aims to gradually raise standards to the high standards already being achieved by the better companies within the UK.

There are currently 678 companies registered with BASIS who, between them, have some 973 pesticide stores registered.

3 REGISTRATION BOARD AND GENERAL ADMINISTRATION

BASIS is administered by a Board whose chairman and vice-chairman are completely independent of the industry but who are recognised within the industry as leaders in their respective fields. The Ministry of Agriculture, Fisheries and Food is involved in the appointment of these persons.

The Board is made up of representatives nominated by the following organisations:

> Association of Independent Crop Consultants (AICC)
> British Agrochemicals Association (BAA)
> Federation of Agricultural Co-operatives (FAC)
> National Association of Agricultural Contractors (NAAC)
> National Farmers Union (NFU)
> United Kingdom Agrochemical Supply Trade Association (UKASTA)

Together with:

> 3 Elected Registered Distributors
> Observers: Ministry of Agriculture, Fisheries & Food (MAFF)
>   Health & Safety Executive (HSE)
>   County Councils' Representative

The Board is serviced by four Committees, the Distribution Committee (Buildings and Transport), Education and Training Committee, Professional Matters Committee and the Fertiliser Advisers' Certification and Training Scheme (FACTS) Committee. Both Board and Committee members receive no remuneration. The working Committees are constituted from the relevant trade associations connected with the pesticide industry and other interested organisations and Government departments.

BASIS is administered by five full-time members of staff from an office central to the UK, and ten part-time, self-employed assessors based throughout the country.

BASIS is financed almost totally from annual registration fees paid by registered distributors; this is currently £175 plus £90.00 for each additional store.   A small amount of income is generated from training activities but the aim is to offer the industry a reasonably priced training and testing service.   The cost of running the scheme is approximately £220,000 per annum.

4 LEGISLATION

The original BASIS scheme agreement was not held in high regard by the EU and subsequently UK Government passed the Food and Environment Protection Act in 1985.   This Act covers all aspects of pesticides from manufacture through to the end user.   As BASIS had had several years in the training and certification field, and also in assessment of stores and transport, the Government looked to BASIS as the blueprint for some aspects of legislation.   All BASIS certificates for storekeepers and technical staff are now requirements under the Control of Pesticides Regulations 1986 and the Code of Practice for Suppliers of Pesticides to agriculture, horticulture and forestry 1990 spells out the requirement for an annual certificate of assessment of stores and staff, all of which actively involves BASIS.   The whole industry welcomed the legislation;   as with many other countries in the world the pesticide industry is under tremendous pressure from environmental organisations.
    FEPA 1985 also covers certification of spray operators.   However this is under the control of the National Proficiency Tests Council.

5 TRAINING

5.1 BASIS

    Since 1981 BASIS has been involved in two sectors of training within the pesticide industry;   Certificates of Competence for storekeepers and Certificates for field sales and technical staff.   BASIS provides certification for people in   agriculture, commercial horticulture, amenity horticulture, forestry, aquatics, seed treaters and sellers.
    The following certificates have been recognised by Ministers as meeting the requirements under the Control of Pesticides Regulations 1986.

        The BASIS Certificate in Crop Protection
        The BASIS Certificate in Crop Protection (Horticulture)
        The BASIS Certificate in Crop Protection (Amenity Horticulture)
        The BASIS Certificate in Crop Protection (Seed Treatment)
        The BASIS Certificate in Crop Protection (Seed Sales)
        The BASIS Certificate in Crop Protection (Forestry)
        The BASIS Certificate in Crop Protection (Aquatic)
        The BASIS Certificate in Crop Protection (Field Vegetables)

        The Certificate of Competence for Nominated Storekeepers
        (Agriculture)
        The Certificate of Competence for Nominated Storekeepers
        (Amenity Horticulture)

5.2 Sales and Advisory Staff Training

    All training is based on eight modules covering 32 days of training:
Growth and development of crops, and commercial production

Recognition and causes of crop disorders
Recognition, biology and control of weeds
Recognition, biology and control of pests
Recognition, biology and control of diseases
Composition, activity and persistence of crop protection chemicals, and biological agents
Application of crop protection chemicals
Safe use, handling, transport and storage of crop protection chemicals.

There is a provision for BASIS training courses at colleges throughout the country, this is not a requirement of the Scheme. Distributors may choose how they train their staff, whether it be in-house or with an external training agency. The only requirement is that they take the BASIS examination at the end of their training.

Candidates are required to sit a multi-choice question paper, submit a project of their own choice relating to the pesticide industry, take an identification test of weeds, pests and diseases and are assessed at four individual field stations on a one to one basis with examiners drawn from the trade and, finally, are interviewed by a viva panel of examiners. The Certificates are only issued after thorough examination; there has been a consistent failure rate of between 14 - 20% in the examinations.

In 1992 BASIS, in collaboration with the industry, launched a Professional Register of staff who are not only qualified to the legal minimum standard as required by FEPA 1985 but are also continually updated through Continuing Professional Development (CPD). Membership is annual and members have an annual certificate and identity card and can use the designatory letters MRPPA (Member of the Register of Practitioners for Pesticide Advice) after their name.

5.3 Nominated Storekeeper Training

The Certificate of Competence for Storekeepers was established by BASIS in 1982. Two certificates are available, one for agriculture/horticulture and one for amenity horticulture.

The course enables people who are operating, and are responsible for, pesticide stores to be competent in all aspects of good housekeeping within the store as it relates to the Code of Practice concerning the storage and transport of pesticides for sale and supply. It is a requirement of the Control of Pesticides Regulations 1986 that there should be one qualified person at each pesticide store.

The course is very practical and lasts for one and a half days and is run on distributors' premises using an actual pesticide store to put over the practical information. Assessment takes the form of a straightforward question and answer paper and concludes with an individual assessment by an examiner.

BASIS organises and administers the storekeeper courses whereas the technical certificate is offered by selected agricultural colleges.

6 STORAGE AND TRANSPORT ASSESSMENTS

6.1 Staff Audit

An annual audit of staff has been made since 1982 throughout the companies registered with BASIS, monitoring stores, the storekeepers and technical staffs. Advice is given to distributors where they fall short of the requirements. This is obviously more important now that certification is a requirement not only for BASIS standards but also of legislation.

6.2 Assessment of Premises

BASIS has been involved in the assessment of pesticide stores since 1979. At the initial assessment of pesticide stores, the practices were such that only 0.5% of stores throughout the UK met the standards then laid down by the BASIS Registration Board.

Registered stores have been assessed annually since 1986; prior to that stores were assessed every 2 or 3 years.

BASIS has a team of ten part-time, self-employed assessors who are based throughout the UK and Northern Ireland, and whose work is co-ordinated by the Technical Executive Officer based at the central office. They are people who have had a lifetime of experience within the pesticide industry, either at the manufacturing or distributing level, and are therefore very conversant with all aspects of pesticide handling. Prior to carrying out their role they are given comprehensive training by BASIS.

Since 1979 there has been a gradual improvement of agreed storage standards as a result of practical experience in the operation of pesticide stores. The major improvement has been the containment of agrochemical spillage and contaminated fire water on storage premises, which had not been considered prior to BASIS. In agreement with the Ministry of Agriculture, Fisheries and Food, stores are now assessed annually because some approved stores were found on subsequent assessment to be below standard, either through changes of staff or the re-siting of a pesticide store within the distributor's premises.

Standards which assessments are made against are:

suitability of site;
adequate capacity and segregation;
soundly constructed of fire resistant materials;
provided with suitable access and exits;
capable of containing spillage and leakage;
dry and frost free;
well lit and ventilated;
marked with appropriate warning signs and secure against theft and vandalism;
equipped, organised and staffed to accommodate intended contents.

A company Registration Certificate is issued annually for each store. This indicates that the store has been assessed and has met the recognised standards, and that the staff has also been audited. Stores which do not meet assessment standards are given one month in which to raise the standards to the required level. Progress is monitored by the BASIS central office.

7 PROGRESS TO DATE

7.1 Storage Standards

Since 1978 BASIS has set and raised standards throughout the UK. Storage standards have been raised from a point where only 0.5% of the industry met the required standards, to the position in 1994 where 41.4% reached the standards at the time of assessment, 55.0% had some faults which were rectified within a month and 3.6% failed the assessment, again being given one month to reach the agreed standards. The reasons for the latter percentage of failure, as previously indicated, are possible changes of staff, re-siting of pesticide stores, new stores, or that the company is newly registered. These figures justify the decision to audit stores annually and indicate how rapidly standards can deteriorate even under this regime.

The standards of storage which BASIS had been implementing since 1978 were taken into account when the Code of Practice for sale and supply was drawn up by the Government, BASIS having been a party to the consultations.

7.2 Certification of Staff

Since 1981, and up to the end of 1994, some 12,256 people have qualified under the Field Sales and Technical Staff category for Crop Protection, 650 under Seed Treatment and 6,371 as store keepers.

Some of these certificates were based on previous experience of staff, applications for which had all been monitored by BASIS. Since 1981 it has been a requirement for all new staff entering the industry to be certificated by examination within three years of entering the industry.

With the announcement of legislation many organisations accepted the BASIS standard and this has now been made a requirement for the Ministry of Agricultures' Agricultural Development and Advisory Service, the Association of Independent Crop Consultants, British Sugar and other independent advisers, and organisations handling amenity products, including County and District Councils. The British Agrochemicals Association, the manufacturers' trade association, requires that its members' staff all take the BASIS examination. BASIS has been involved in discussions with many countries throughout the world but it would appear that there are no similar schemes operating, other than now in Australia and Ireland following visits to BASIS by industry representatives.

The BASIS Board, having full support from both MAFF and HSE, agreed to open its doors not only to organisations with an agricultural interest in pesticides but also to those who have a non-agricultural pesticides interest.

8 CONCLUSION

BASIS to date has achieved a degree of recognition throughout the UK and the rest of the world. However, neither BASIS nor the industry can be complacent. It would only need one major fire incident in a pesticide store, resulting in considerable water pollution, to undo much of the good work which has been established over the past fourteen years. The industry is well aware of this situation.

Pesticide Loading to the North Sea: the Red List Reductions

M. J. Pearson

NATIONAL RIVERS AUTHORITY, ANGLIAN REGION, KINGFISHER HOUSE, ORTON GOLDHAY,
PETERBOROUGH PE2 5ZR, UK

1 INTRODUCTION

Historically, there has been much concern about the quality of the North Sea. Steps were first taken in 1974 to formulate policy to eliminate or reduce pollution. In essence, the countries bordering the North Sea adopted a Convention, under which they agreed to monitor and control the discharge of hazardous substances and nutrients to protect the marine environment. A number of Ministerial Conferences were held at which Declarations and Agreements were signed, binding the UK Government to meet certain targets for pollution load reductions for "listed" substances, over a specified time period. These substances were selected for priority control due to their toxicity, persistence in the environment and potential to accumulate in marine life. This list was originally known as the Red List, but in 1991 was extended and referred to as Annex 1A. The list contains metals, organometals, pesticides and organic solvents. Since 1989, in England and Wales, the National Rivers Authority (NRA) has been responsible for monitoring the inputs of these substances to the aquatic environment.

2 INTERNATIONAL OBLIGATIONS

The UK is required through various international agreements and commitments to reduce the quantities of certain hazardous substances and nutrients entering the sea. The Convention for the Prevention of Marine Pollution from Land-Based Sources (the Paris Convention), was adopted on 4 June 1974 and brought into effect on 6 May 1978. The Convention is administered by the Paris Commission (PARCOM). By 1978 it was generally perceived (by the public and also by some Governments) that slow progress was being made in reducing pollution of the North Sea. As a result, a series of Ministerial Conferences were held attended by the Environmental Ministers of all North Sea riparian countries. The first was held in Germany in 1984; the second in London in 1987; the third at the Hague in 1990; and a fourth is planned in Denmark for 1995. At the end of each Conference the Ministers from the participating countries agree objectives by way of Declarations. These Declarations are not legally binding and it is up to each Government to decide how to achieve the stated objectives.

In the Declaration from the Second Conference in 1987, the need to adopt a precautionary approach in relation to the most dangerous substances (defined as those that are persistent, toxic and liable to bioaccumulate) was a major feature. Amongst other issues the Ministers agreed to take measures to:-

- reduce the input loads of dangerous substances, from rivers and estuaries, to the North Sea by around 50% by 1995 using input load in 1985 as a baseline;
- reduce inputs of phosphorus and nitrogen by around 50% between 1985 and 1995 into areas where these inputs are likely to cause pollution.

For the dangerous substances, each country had to draw up its own priority list from a reference list of 170 substances. In the UK a list of 23 substances was created based on toxicity, persistence, bioaccumulation and volume of usage: this list is commonly referred to as the Red List. Inputs from point sources (e.g. industrial discharges) were to be reduced through the application of best available technology (BAT). Diffuse source inputs (e.g. inputs from run-off of pesticides from agricultural land) were to be reduced by controls on supply, use and safe disposal of products.

At the Third Conference in 1990, Ministers agreed to extend the Red List and produced a common list of 36 dangerous substances, referred to as the North Sea Conference Common, or Annex 1A, List (see Table 1). The sources and uses of substances included in the UK Annex 1A list are shown in Table 2. Ministers also reiterated their previous commitments concerning inputs of hazardous substances and agreed to:-

- achieve significant reductions of these 36 substances, from rivers and estuaries, to the North Sea by around 50% between 1985 and 1995;
- to reduce the total inputs, from all sources, of dioxins, mercury, cadmium and lead by around 70% or more between 1985 and 1995 (provided that the use of best available technology or other low waste technologies enable such reductions);
- to reduce the input of nutrients, by around 50%, between 1985 and 1995 in areas where these inputs are likely to cause pollution;
- to make substantial reductions in the quantities of pesticides reaching the North Sea with special attention to phasing out those which are the most persistent, toxic and liable to bioaccumulate:
- to phase out and destroy all identifiable PCBs by 1999.

The area of sea affected by the North Sea Declaration is confined to the North Sea and Channel, and immediately adjoining waters connecting it to the Baltic. The Government set out its intentions with regard to these commitments in a guidance note issued after the 1990 North Sea Conference. One point from the note is that the Government is applying the policies to all UK coastal waters and not just the North Sea.

The UK Government announced in 1990 that a programme to monitor riverine inputs to the sea was to be introduced. The objective of the programme was to monitor on an annual basis at least 90% of all inputs of each substance via rivers at tidal limits, and via sewage effluent or direct industrial discharges downstream of tidal limits. The NRA is responsible for implementing the monitoring programme.

3 THE CALCULATION OF LOADS

The international agreements relating to the reduction in hazardous substances and nutrients discharged to the North Sea refer to load reductions. Loads are the product of the contaminant concentration and the river or effluent flow. PARCOM has provided standard methods for the estimation and calculation of input loads to coastal waters, and the NRA has adopted these recommendations in reporting the results of its sampling programme for Annex 1A purposes.

The NRA's monitoring programme for North Sea purposes consists of about 300 sampling sites. All the main English and Welsh river systems are generally sampled monthly at a sampling point close to, but upstream of, the tidal limit for a wide range of contaminants. In addition, all major direct discharges of trade and sewage effluent entering downstream of that sampling point are sampled, as are all major coastal discharges. It is impossible to sample 100% of the total input load, as the final few % of that load are inevitably spread in small amounts across large numbers of very small streams or effluents. The PARCOM recommendations allow for this and suggest the aim should be to sample 90% of the total load.

PARCOM gives two methods for calculating river loads: one based on mean annual rates (for which river flow on each day of the year is needed), and one using average loads on the sampling day (for which river flow is only needed on each sampling day). The NRA uses the second method, as data on annual mean river flows are not readily available. Effluent loads are calculated in a similar way.

As part of the PARCOM standard methods, limits of detection (LOD) for the chemical analysis of certain contaminants are given. The LOD is the break point at which any analytical method can just detect the presence of the contaminant in the sample. Any amount smaller than this limit is indistinguishable from a zero amount and cannot be quantified. However, rather than recording all such amounts as zero (which would not be true), it is usual to refer them as being less than the numerical value of the LOD e.g. <10 ng/l,<5μg/l etc.

Two sets of load estimates are calculated. The first treats results recorded as less than the LOD as having a true concentration of zero. When this concentration is multiplied by a flow, a low load value is obtained. The second set treats such results as having a true concentration at the LOD. When this concentration is multiplied by a flow, a high load is obtained.

Where load discharged to a river catchment or to a coastal sea zone is to be quoted the question arises whether the high or low load should be used. There is no right answer as neither load is wholly correct. However, any concern which may be at the size of a low load has to be reinforced by the knowledge that the true load will be higher still. Thus, there is some advantage in looking at low loads first and this is the NRA's usual course of action. Ultimately, it is only by looking at both loads and at how much of the concentration data are above the LOD that understanding can be as complete as possible.

The data collected by the NRA for Annex 1A purposes since 1991, can be used to build up a picture of total load of a particular contaminant discharged in England and Wales. Data collected are processed to give:-

· loads for individual estuaries (summation of river, sewage and industrial loads).
· loads for coastal sea areas (North Sea, English Channel, Bristol Channel and Irish Sea).
· loads for North Sea (North Sea and English Channel).
· loads for England and Wales.

The input loads can then be ranked in order of size and a national load reduction programme implemented to target specific discharges.

4 LOAD REDUCTION TARGETS

One of the main objectives of the North Sea Declaration is the commitment to reduce input loads of dangerous substances to the North Sea by around 50% between 1985 and 1995. In order to assess whether, or not, this target has been achieved, a 1985 baseline load of each contaminant is required. Such baseline data, however, only exist for 6 substances: cadmium, mercury, copper, zinc, lead and gamma HCH (lindane). These 1985 baselines were published by the DoE in its UK North Sea Action Plan. The remaining 30 Annex 1A substances do not, therefore, have 1985 baseline data. Since 1991, the NRA has been collecting data annually for these substances, and an artificial baseline could be constructed using either 1991 data or 1991/1992 data. However, this may not be straightforward as the data show extreme variability due to many of the analytical results being at, or below, the LOD. The high and low load figures can differ markedly, making an assessment of a baseline load very difficult.

5 RESULTS OF MONITORING

The levels of contaminant inputs monitored over the period 1990-1993 varied around the coastline; the North Sea and Irish Sea generally receiving the highest loads of many substances. The monitoring programme has also provided some important information on patterns and trends in inputs of contaminants. It is clear that the relative contributions from different sources varies between different contaminants according to their pattern of use. For example, the main source of some metals (mercury, cadmium, arsenic) is direct industrial discharges, whereas other metals (copper, lead, zinc) originate from diffuse sources within the river catchment. Most organochlorine (e.g. aldrin) and organophosphorus (e.g. fenitrothion) pesticides on the priority list are, or have been, insecticides used in the UK. The others (e.g. atrazine, simazine, trifluralin) are herbicides. The application of pesticides to land can give rise to diffuse pollution, e.g. from spray drift, leaching, surface run-off and farm waste. Additionally, inputs of pesticides can arise from point sources (e.g. pentachlorophenol and lindane, both used in wood treatment processes).

Considering the overall inputs to the sea around the whole of the England and Wales coastline, there has been significant reductions in the loads of many of the Annex 1A substances. Of the six substances with published 1985 baselines, mercury, cadmium, copper and gamma HCH (Lindane) have achieved the 1995 load reduction target (50% of the 1985 load). Zinc and lead, however, have not been reduced significantly over this period. Figure 1 shows the annual loads of these 6 substances, together with their 1985 baselines, and 1995 targets. These are high loads, as these were the only data published by DoE for 1985. The reductions achieved for the

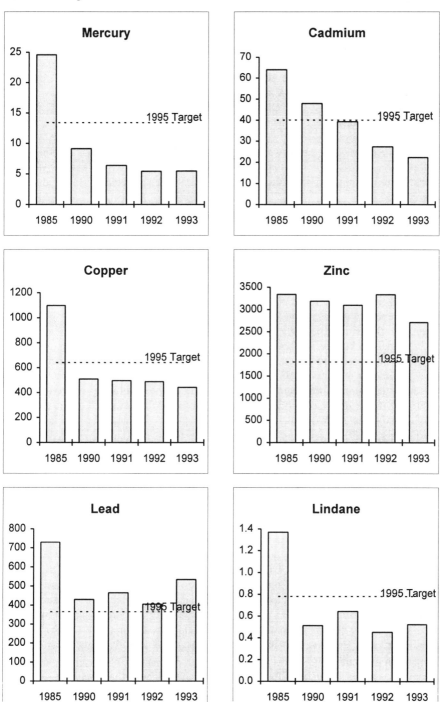

Figure 1 *Annex 1A High Load Metals & Lindane (Tonnes/Yr)*

metals are a combination of legal controls of direct discharges by the NRA via consents, the investment programmes of discharges, and the effects of the recession.

In contrast, there are no baseline data, or load reduction targets, for pesticides. The NRA has, therefore, adopted the policy of calculating and assessing pollution reduction achievements using low load data. Figure 2 gives examples of 6 Annex 1A pesticides. Generally, over the period 1990-1993, pesticide loads to the seas around England and Wales have gradually reduced, although some variation exists from year to year. The reductions achieved for pesticides are generally due to controls on manufacture and usage.

The above data are forwarded to DoE on an annual basis, and together with information from other regulatory organisations (e.g. HMIP, MAFF, RPBs), manufacturing industries and users, the Government can assess overall load reductions for the UK.

6 NRA ACTIONS

6.1 Ranking Individual Loads to Target Major Sources

For the North Sea Conference Surveys, and for both high and low load data on dangerous substances, the importance of individual inputs is assessed by the NRA. The percentage that each input contributes to the national total load is calculated and inputs are then ranked in order of size.

Such ranking has to be interpreted with some care. In those cases where an individual load is based on ten or more positive results and on ten or more measured flow recorder results, the accuracy of the data is relatively high. However, where a more limited number of the samples analysed through the year have produced positive values, or where flow has had to be estimated rather than measured, then the accuracy of the load calculation is lower.

In spite of the caution required when interpreting the data, the NRA believes ranking individual loads may be a useful tool for indentifying the major sources of input of individual substances and is investigating further.

6.2 Reviewing Consents

Before any effluent can be discharged, whether industrial or sewage, to a river, estuary or direct to sea it needs the written consent of the NRA. Where dangerous substances are known to be present in the effluent in significant amounts, the consent will normally identify those substances and impose a concentration limit and a restriction on volume. Consequently, a limit on the maximum load that can be discharged is imposed on the discharger. Where a specific catchment is found to be a major source of a particular substance then discharge consents containing that substance will be reviewed. While the individual cases will each have to be judged on their merits, the overall objective is to reduce the permitted load that can be discharged from the catchment.

The introduction of Integrated Pollution Control (IPC) means that some discharges containing dangerous substances will be authorised under the Environmental Protection Act 1990, by Her Majesty's Inspectorate of Pollution (HMIP). In these cases the condition on the authorisation will be set at levels at least as stringent as those required by the NRA. In formulating the NRA requirements,

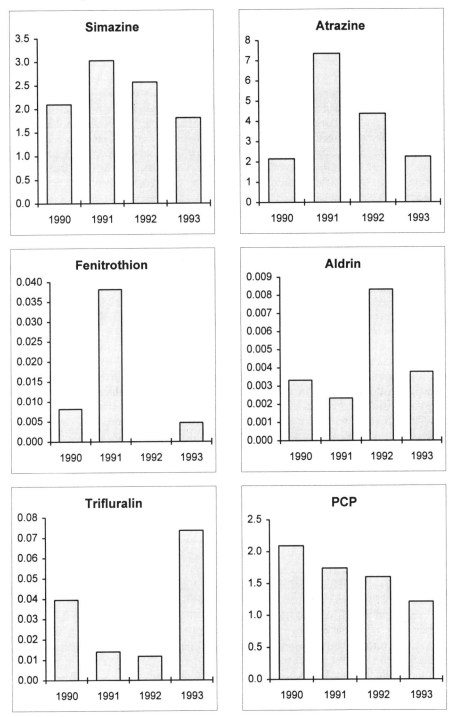

Figure 2 *Annex 1A Low Load Pesticides (Tonnes/Yr)*

the same approach will be adopted, i.e. an agreed reduction in the load discharged will be set.

6.3 Surveys of Freshwater Catchments

Some freshwater rivers are a major source of certain substances. In such cases, an assessment is required to determine whether the presence of the substance in the river is the result of point or diffuse inputs. This involves the identification of key sources of the substance in the catchment and an evaluation of their contributions to the river system. Typically, such an assessment will commence with an examination of the consents for effluent discharges within the catchment. This will identify any point sources of the substance under investigation. Indeed, in some cases this initial examination of the available data will indicate that most of the substance load is introduced by effluents and a review of the discharge consents will be required to achieve the necessary load reductions.

However, if it is clear from a review of the upstream discharges, that the major source of a substance is from diffuse inputs, it becomes difficult for the NRA to achieve load reductions. The methods normally employed for reducing diffuse inputs of substances, such as pesticides, are to alter the way in which the substance is used, to encourage a replacement with a more "environmentally friendly" version, or to press for a complete ban of a particular substance.

Where diffuse inputs are the major source of a substance in a river, the NRA tries to ensure that river monitoring is comprehensive enough to identify particular areas of the catchment which appear to be the main sources of contamination. Control mechanisms available to the NRA will almost certainly need to involve Central Government (e.g. MAFF) in any initiative to reduce or change the use of a particular substance, such as a pesticide. Comprehensive monitoring will need to be continued on the upstream freshwater river for some years to confirm that loads have been reduced.

6.4 Pesticide Initiatives

There are 18 pesticides on the Annex 1A list. With the exception of pentachlorophenol (PCP) and some of the organochlorines, the majority of the load derived from these compounds comes from diffuse run-off i.e. from the land rather than discharges from industry or domestic sources. This diffuse pollution can result either from the use of pesticides to protect agricultural crops or from weed control in non-agricultural situations.

Diffuse source run-off of pesticides is difficult to control. However, it is sometimes possible to withdraw a pesticide from use in situations where the run-off is significant; this method has been adopted for the organochlorines and for the non agricultural use of the triazine herbicides, atrazine and simazine. Alternatively, steps can be taken to reduce the overall use of the pesticide. And even where there is no reduction in use, it may be possible to improve the way they are used so that run off is minimised.

The NRA's strategy for reducing pesticide pollution includes measures to control point source discharges directly by using powers under the Water Resources Act (1991) and to influence pesticide use patterns to reduce pollution from diffuse sources. Where the NRA's monitoring programmes indicate that pesticides are present in the aquatic environment in significant quantities the evidence is passed to the relevant Government Departments. They can then address the problem during reviews of approval for use. These reviews are carried out under the Food and Environment Protection Act (1985) at present, but in future it will be through implementation of the EC "Uniform Principles" Directive. These reviews may result in the restriction of use of the pesticide or the selective withdrawal of approval under those situations where it is known to be causing a particular problem or, in extreme cases, the complete withdrawal of the approval for use.

Of equal importance in the NRA's strategy is the encouragement of Best Practice in the manufacture, transport, storage, use and disposal of pesticides. This has included the significant inputs to statutory and non-statutory guidance on the use, storage and disposal of pesticides, particularly in relation to the measures included for pollution prevention. This aspect is seen to be particularly important in reducing the number of small incidents resulting from use and disposal. In addition, the NRA has produced its own guidance leaflet for farmers to increase awareness of the potential problems which can be caused by pesticides in water.

7 THE FUTURE

The Fourth International Conference on the Protection of the North Sea will be held in June, 1995 in Denmark, and the Declaration will be signed. A Declaration is a collection of political statements and the NRA awaits the DoE's guidance note which will set out the current thinking of Government about the implications of each Section of the 4th Declaration, and will indicate what action is expected, by whom and to what timescale. The NRA will then work towards meeting its commitments under these new international obligations.

Table 1 *Annex 1A Substances*

Mercury	DDT	Fenitrothion
Cadmium	Drins	Fenthion
Copper	Trifluralin	Malathion
Zinc	Trichlorobenzene	Parathion
Lead	Trichloroethylene	Parathion-methyl
Tributyltin	Tetrachloroethylene	Azinphos-ethyl
Triphenyltin	Hexachlorobenzene	Azinphos-methyl
Chromium	Hexachlorobutadiene	Atrazine
Nickel	Carbon Tetrachloride	Simazine
Arsenic	Chloroform	Pentachlorophenol
Hexachlorocyclohexane	Endosulphan	1,2 Dichloroethane
Dichlorvos	Trichloroethane	Dioxins*

* Dioxins will be monitored by the DoE.

Table 2 *Sources and Uses of Annex 1A Substances*

SUBSTANCE	SOURCE	USES
METALS		
Arsenic	Point and diffuse	Wood preservation, manufacture of glass, alloys, medicines and semi-conductors, by-product of smelting industry.
Copper	Point and diffuse	Metal plating industry, agriculture, manufacture of alloys, copper wire & piping, textile dyeing, glass & ceramics, catalyst in vinyl chloride production, manufacture of wood preservatives, rayon, paint pigments, active ingredient in marine antifouling paint.
Zinc	Point and diffuse	Manufacture of alloys, electroplating and galvanising, manufacture of rayon textiles, production of paper, fungicides, rubber, paint, ceramics, glass, reprographic materials, hygiene products.
Cadmium	Point and diffuse	Manufacture of pigments & stabilisers, batteries, cement, agrochemicals, alloys, solders, photoelectric cells, electrodes, electroplating, photographic processes, deoxidiser.
Mercury	Point and diffuse	Manufacture of batteries, agrochemicals, pharmaceuticals, mirrors, thermometers, barometers, chlor-alkali production, catalysts, dentistry.
Lead	Point and diffuse	Manufacture of batteries, production of anti-knock agents for petrol, cable covering, solder, pigments, type-metal, building material, radiation shields, cement manufacture, shot for shooting and fishing.
Nickel	Point and diffuse	Metal plating industry, iron and steel production.
Chromium	Point and diffuse	Metal plating industry, iron and steel production, pigment production, textile colouring, lithographic and photographic applications, glass manufacture, ceramics production, leather tanning.

Table 2 (continued)

SUBSTANCE	**SOURCE**	**USES**

ORGANOPHOSPHORUS PESTICIDES

Parathion	Mainly diffuse	Agriculture (never approved in the UK)
Parathion-methyl	Mainly diffuse	Agriculture (never approved in the UK)
Azinphos-methyl	Mainly diffuse	Agriculture (withdrawn in the 1990's in the UK)
Azinphos-ethyl	Mainly diffuse	Agriculture (withdrawn in the 1980's in the UK)
Fenthion	Mainly diffuse	Agriculture and veterinary (never approved in the UK)
Fenitrothion	Mainly diffuse	Agriculture, public health and domestic.
Dichlorvos	Mainly diffuse	Agriculture, public health and domestic.
Malathion	Mainly diffuse	Agriculture and domestic

ORGANOCHLORINE COMPOUNDS

Carbon tetrachloride	Point and diffuse	Petroleum refining/coal processing, halogenation of non-aromatics, fabric mills, paper and pulp mills, manufacture of plastics, synthetic rubber, pharmaceuticals, flavourings, perfumes, cosmetics, industrial organic chemicals, electronic components.
Hexachlorobenzene	Mainly point	Biocide and chemical synthesis
Hexachlorobutadiene	Point	Solvent, refrigeration systems, hydraulic systems, transformer oil, dielectric fluid.
Polychlorinated Biphenyls (PCBs)	Point and diffuse	Heat exchange agent, dielectric fluid, lubricant.
1,2-Dichloroethane	Point	Intermediate in the production of chlorinated hydrocarbons e.g. vinyl chloride, 1,1,1-trichloroethane, trichloroethylene and tetrachloroethylene, solvent.
1,2,3 -Trichlorobenzene	Point	Solvent and chemical synthesis
Tetrachloroethylene	Point and diffuse	Solvent and chemical synthesis
Trichloroethylene	Point and diffuse	Solvent and chemical synthesis
Trichloroethane	Point and diffuse	Solvent
Chloroform	Point and diffuse	Solvent and chemical synthesis

ORGANOCHLORINE PESTICIDES

Aldrin	Mainly diffuse	Agriculture (use prohibited 1989)
Endrin	Mainly diffuse	Agriculture (use prohibited 1984)

172

Table 2 (continued)

SUBSTANCE	SOURCE	USES
Dieldrin	Point and diffuse	Agriculture, industrial & domestic (use prohibited 1989)
Endosulfan	Mainly diffuse	Agriculture and forestry (use severely restricted)
DDT	Mainly point	Use as insecticide prohibited in the EU (UK from 1984)
Lindane (Gamma-HCH)	Mainly diffuse	Agriculture, insecticide and public health
Pentachlorophenol	Point and diffuse	Wood preservative

OTHER PESTICIDES

Atrazine	Mainly diffuse	Agriculture (use prohibited for non-agricultural use 1993)
Simazine	Mainly diffuse	Agriculture (use prohibited for non-agricultural use 1993)
Trifluralin	Mainly diffuse	Agriculture

NUTRIENTS

Ammonia (NH_3)	Point and diffuse	Domestic and agriculture
Nitrate (NO_3)	Point and diffuse	Domestic and agriculture
Nitrite (NO_2)	Point and diffuse	Domestic and agriculture
Total Oxidised Nitrogen	Point and diffuse	Domestic and agriculture
Total Nitrogen	Point and diffuse	Domestic and agriculture
Total Phosphorus	Point and diffuse	Domestic and agriculture

ORGANOMETALS

Tributyltin	Mainly diffuse	Antifouling paints, wood preservation
Triphenyltin	Mainly diffuse	Used in the synthesis of biocides
Total Organic Tin	Mainly diffuse	Antifouling, wood preservation, biocides

Subject Index

Grain, 67
Granular activated carbon (GAC), 126,
 130
 bed life, 127
Granule application, 59
Grassland, 2
Groundwater protection, 101
 contamination, 112, 114
 in Europe, 115, 116
 in USA, 115
Groundwater Ubiquity Score (GUS),
 118
Growth regulators, 3

Handlers, 45
Hardy nursery stock, 2
Harmonised Monitoring Programme,
 105
Hazards, 143
Health and Safety at Work Act, 29
Health and Safety Executive, 45
Health Guideline Value, 38
Heptachlor, 91, 113
Her Majesty's Inspectorate of Pollution
 (HMIP)
Herbicides, 79, 109
 phenyl urea, 109
Hexachlorobenzene, 69, 91, 103, 106,
 113
High Performance Liquid
 Chromatography (HPLC), 81, 108
Hops, 2
Hormore weedkillers, 9, 13
Hospitals, 21
Humic materials, 129
Hydraulic nozzles, 55, 147
Hydrogen peroxide, 129, 140

Imazypyr, 9, 13
Induction hoppers, 30
Industrial sites, 8
Ingestion, 45
Inhalation, 44, 45, 52
 exposure, 32
Injection system, 58
Insecticide usage on cereal crops, 4
Inspirable fraction, 36
Integrated pest management (IPM), 19

Integrated Pollution Control (IPC),
 166
Ion trap GC-MS, 6
Iprodione, 5
Isle of Man, 87
Isle of Wight, 109
Isoproturon, 14, 104, 107-109, 116,
 124, 125, 128

Karate, 24
Keds, 43, 87
Knapsack sprayers, 56

Lamda cyhalothrin, 20, 22-24
Landscape design, 13
Laser light diffraction, 55
Lice, 21, 87
 salmon, 136
Limit of determination, 5, 163
Lindane, 43, 67, 69, 87, 91, 97-103,
 106, 113, 116, 124, 125, 144,
 164, 169
Linuron, 116
Liquid chromatography-mass
 spectromertry, 6
Load reductions, 163
Lough Neagh, 88
Low level induction bowls, 58
Lure and kill traps, 21

Malathion, 67, 91, 103, 106,169
Maneb, 104
Marine fish farming, 135
Mathematical modelling, 139
Maximum Acceptable
 Concentration (MAC), 91, 122, 124
Maximum Exposure Limit (MEL), 29
Maximum Residue Level (MRL), 5
 63, 70, 137
MCPA, 13, 104, 107
Mecoprop, 10, 104, 107-109, 116,
 124, 125
Member of the Register of
 Practitioners for Pesticides
 Advice, 158
Metabolic pathways for pesticides,
 43
Metabolites, 5